52 RANDOM WEEKEND PROJECTS

All experiments included in this book are provided as a resource for readers. We recommend that an adult supervise children at all times when experiments are performed. Experiments are not recommended for children under thirteen years of age. Neither the author nor the publisher accepts any responsibility for any loss, injury, or damages sustained by anyone resulting from the experiments contained in this book.

52 RANDOM WEEKEND PROJECTS. Copyright © 2020 by MDM Productions. All rights reserved. Printed in the United States of America. For information, address St. Martin's Publishing Group, 120 Broadway, New York, NY 10271.

www.stmartins.com

Design by Jonathan Bennett, with Susan Walsh

The Library of Congress Cataloging-in-Publication Data is available upon request.

ISBN 978-1-250-18450-4 (trade paperback)
ISBN 978-1-250-18451-1 (ebook)

Our books may be purchased in bulk for promotional, educational, or business use. Please contact your local bookseller or the Macmillan Corporate and Premium Sales Department at 1-800-221-7945, extension 5442, or by email at MacmillanSpecialMarkets@macmillan.com.

First Edition: March 2020

10 9 8 7 6 5 4 3 2 1

CONTENTS

A NOTE ABOUT SAFETY

Great science happens through accidents. However, we want to make sure that we are safe first and foremost.

These projects should not be attempted without adult supervision.

Misuse, or careless use, of tools and/or projects may result in serious injury or death. Use of this content is at your own risk, and you assume all responsibility for your actions.

Do not point any projectile projects at people or any living thing. Please be aware of populated areas.

Be aware of your surroundings, the wind, and other potential threats like traffic and power lines.

Respect and extreme caution are always required when testing your curiosity and exploring your creativity.

Enjoy!

ACKNOWLEDGMENTS

The King of Random team would like to thank the following:

Our manager, Larry Shapiro, who never let up on this process and kept driving the ball forward.

Byrd Leavell, for guiding us on the big picture from the beginning.

Marc Resnick, Hannah O'Grady, and the team at St. Martin's Press.

Jake and Ritchie from Sonic Dad.

Nicolette Zaretsky, Danielle Colucci, Ryann Stibor, and Chellie Grossman, whose input during this process was invaluable and without whom this book could not have been possible.

And most important, all our fans, for being the reason we've pushed ourselves so hard to get this far.

52 RANDOM WEEKEND PROJECTS

INTRODUCTION

Throughout my entire life, I have always been one to take on challenges, push myself, explore the world, and lean into the sharp points of life.

I have been a roughneck blowing up sides of mountains for oil companies, and I have learned how to fly and pilot for Delta Air Lines. I guess my career has always been to look at things and ask, "How can I do that?" or, "How does that work?" Then I just work hard to figure it out.

I have always asked questions. Being curious is what it is all about. It is how we achieve greatness. Whether you are building a rocket that you launch in your backyard, or launching one into space, it starts with a question and involves problem-solving to figure it out.

In 2009 the economy was struggling and there were rumors of another Great Depression. So with a wife and kids to take care of, I started on a personal journey to intimately understand how important things like lightbulbs, electricity, and filtered water are produced, so I could make them myself if I needed to.

I spent the next year tinkering and experimenting with anything I could get for free from the classified ads, and making discoveries about things that blew my mind, like building arc welders from microwaves, rocket motors from sugar, and high-voltage electricity from glass bottles and salt water.

After I'd shared random experiments with friends and

acquaintances for a few months, people began asking, "Who is this guy?" and "Why does he know these kinds of things?" And during one historic conversation, someone even commented, "Dude, you're like, the king of random."

I realized the things I was experimenting with were peculiar, and people seemed unusually interested to know more, so I decided to learn how to make videos and share them on YouTube.

Growing up, I'd always wanted to be some kind of superhero inventor like Tony Stark who could make anything he could imagine. And that's what it's all about.

Some of my favorite projects have been inspired by reading books and trying these ideas out for myself. My hope is this book provides that same experience for you.

Taking on new challenges has always given me a great sense of accomplishment. I love the feeling I get when figuring things out, solving a problem, and being creative. Whether it is learning how to build a foundry or how to fly a plane, it's the same thought process. You set a goal, you make a plan, and you follow it through.

I hope you're inspired to try something fun, think outside the box, and be the next experimenter who changes the world.

Have fun, be safe, and create something great.

EASY PROJECTS

Shoot a matchstick with power,
blast toothpicks into fruit, and lob
fiery darts over twenty feet!

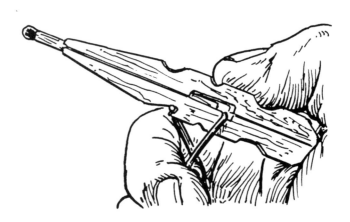

CLOTHESPIN GUN

01

SAFETY KEY:
  + Sharp Objects

SKILL LEVEL:
  EASY
  INTERMEDIATE
  ADVANCED

APPROXIMATE TIME:
  20 minutes

WHAT YOU'LL NEED:
  + Wooden
    clothespin with
    metal spring

  + Wood glue

  + Scrap paper

  + Sharpie

  + Utility knife

  + Wooden matchsticks

  + Fruit (for target practice)

LET'S BEGIN

BUILDING YOUR POCKET PISTOL

1. Push on the clothespin sideways to break the tension, separate the pieces, and remove the spring.

2. Hold the two wooden pieces back to back so the notches line up near the center. Using the Sharpie, mark the top piece about half an inch from the hole. Color from the mark up to the notch, and add an angled marking on the bottom.

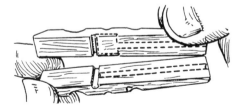

3. Color in the sections of the pin that you'll be cutting away.

4. With your utility knife, carefully carve out all of the areas marked with your Sharpie.

PRO TIP: Don't forget the notch where you marked the angle on the bottom.

Notch carved out more on lower right in diagram

5. Add a dab of wood glue to a piece of scrap paper. Carefully slide the pieces facedown through the glue and press them together. Wipe away any excess glue and let sit for about ten minutes.

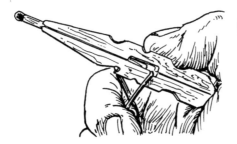

6. Push one end of the spring into the inner chamber, and hook the other end over the outside notch that you carved out following the angled mark. With the spring in place, your pocket gun is finished and ready for action!

TARGET PRACTICE

All we need now is ammunition! Slide a matchstick loosely into the gun barrel. As you push in the match, the gun cocks itself automatically and is instantly ready to fire. You can hold it just like a little pistol, and when you're ready to shoot, simply pull the trigger. It's amazing to see how much power is in the spring!

ALSO TRY: If you're feeling really adventurous, try lighting the matches first, then lobbing the fiery darts high into the air.

TARGET PRACTICE WITH FRUIT: If you break off the tip of the match, it exposes a broader surface, propelling it faster. Test it on an apple; it will impale the apple. The closer you are, the deeper it penetrates.

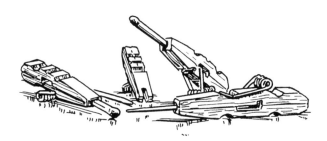

Now you can turn a simple clothespin into a spring-loaded pocket pistol! Experiment with different ways to fire them, from accuracy shots to soaring flames, with this mini weapon of mass satisfaction.

FUN FACT: There were over one hundred US patents issued for clothespins between 1852 and 1857. Clothespins lost their value quickly in 1938 when the first electric clothes dryer was introduced to the public.

Who knew a piece of aluminum foil and a box of matches could be turned into the ultimate desktop weapon? These rockets may be tiny, but they are impressively powerful, leave a cool trail of smoke, and are powered by a single match head!

MATCHBOX ROCKET

02

SAFETY KEY:

+ Fire + Safety gloves + Adult supervision advised

WARNING:

☠ Although these rockets are only fueled by one match head, they do get hot enough to burn fingers and leave scorch marks in carpets.

SKILL LEVEL:

EASY

INTERMEDIATE

ADVANCED

APPROXIMATE TIME:

1 hour

WHAT YOU'LL NEED:

+ Box of wooden matches

+ Package of wooden skewers, approx. 10 inches long

+ Aluminum foil

+ Aluminum foil tape

+ Tealight candle

+ Empty cereal box

+ Pliers

+ Scissors

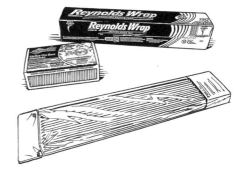

LET'S BEGIN

PREPARING YOUR KIT

1. With a pair of scissors cut the heads off a bundle of matches.

PRO TIP: Cut your match heads into a container lined with a sock so they don't bounce out.

2. Using the template below, transfer the marks from the diagram onto the skewer (it should be about half its original size and fit the length of the matchbox almost exactly). Carefully cut the skewer to length.

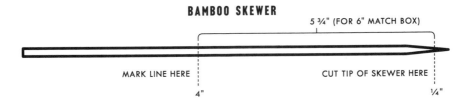

BAMBOO SKEWER

5 ¾" (FOR 6" MATCH BOX)

MARK LINE HERE

4"

CUT TIP OF SKEWER HERE

¼"

3. Tape the body template (page 13) to a piece of paperboard that you can cut from the cereal box. This will be your stencil, so cut the edges out as cleanly as possible.

4. Use the utility knife to cut out the little square on the template (page 13). Before using the template, make sure that its length and width are the size of your matchbox. Trace the square onto a strip of aluminum tape. Each will make one set of rocket fins, so make as many as you'd like.

5. Once cut out, fold them "point to point" in both directions. Pinch them at the base and push your fingers together, so that when you crease them down it looks like a little X-wing. Snip off the point.

> **PRO TIP:** Store your match heads and rocket fins in a homemade soda cap container (see chapter 10).

6. Lay a paper towel on top of a sheet of aluminum foil and then carefully fold the stack three times, making it four layers deep and just a bit larger than the cardboard template. Trace the template onto your foil stack and cut it out. This should give you four identical rocket bodies.

> **HELPFUL HINT:** Normally the foil would stick together, but the paper towel solves that problem. With this system, you can cut out dozens of rocket bodies in only a few minutes.

7. Poke a small hole in the top of the matchbox, about half an inch from the end, and you're done! The template and the skewer were designed to fit perfectly inside the matchbox. Toss in your rocket fins, spare matches and match heads, and tea candle, and you've created a portable assembly station that you can take just about anywhere.

ASSEMBLING YOUR ROCKET

LET'S GET TO WORK AND BUILD SOME ROCKETS. THE FINISHED ROCKET IS LIGHT AS A FEATHER, BUT SURPRISINGLY STABLE IN FLIGHT.

1. There are two markings on the template indicating where to place the top of the skewer and the top of the match head. Place a skewer and match head on an aluminum rocket body. Tightly roll the aluminum up, pinching the tube right above the match head at the end. Crimp the tip with a pair of pliers.

2. Attach the rocket fins by peeling off the paper covering the sticky stuff on the back, then pushing the rocket body through the hole in the center. Pinch the four fins until they stick firmly in place near the bottom of the rocket. Slide your rocket off the skewer and you're almost ready for liftoff!

1. Load your rocket by pushing the tapered end of the skewer into the aluminum rocket. Twist it upward until the skewer touches the match head inside the rocket.

2. Push the skewer through the hole in the matchbox to create your launchpad.

3. Light the tea candle and position the flame just under the tip of the rocket. The foil will quickly warm until the match head reaches its auto-ignition temperature, at which point—bombs away! Your rocket will shoot off with an impressive amount of speed and power, leaving an awesome smoke trail and traveling up to forty feet!

PRO TIP: The rockets will propel the farthest if they're launched from a stable base. Any give on the launchpad will absorb some of the energy and the rocket won't go as far. If you still have problems, make sure your crimp is tight.

As simple as this kit looks, it took over a year of prototyping and playing to finally figure it out. It's not rocket science...or is it?

ALUMINUM FOIL FOR ROCKET BODY

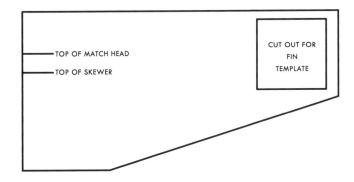

FUN FACT: In the sixth century, the Chinese are thought to have made a type of match by dipping a pine stick in sulfur and letting it dry.

Whether your hero is Bruce Lee or a hero in a half shell, these easy-to-make nunchucks are a great weekend project for you and your band of warriors.

NUNCHUCKS

03

SAFETY KEY:

+ Eye protection

WARNING:

☠ Check the weapons laws in your state. Nunchucks may not be allowed outdoors, or may only be permitted as decoration. These are not toys.

SKILL LEVEL:

EASY

INTERMEDIATE

ADVANCED

APPROXIMATE TIME:

30 minutes

WHAT YOU'LL NEED:

+ 1-inch-thick dowel

+ Handy link chain

+ Electrical tape, black

+ 2 Everbilt #210 eyebolts

+ Sharpie

+ Drill with $^3/_{32}$-inch bit

+ Hacksaw

+ Two pairs of pliers

+ Electrical tape, any color preference

LET'S BEGIN

BUILDING YOUR NUNCHUCKS

1. Using your hacksaw, cut your dowel into two 9-inch pieces.

2. Slowly cover your sticks in electrical tape. Start by putting the tape on at a little bit of an angle so that as you turn the dowel, it overlaps the piece before it. Once you get to the end, wrap a couple of loops around the tip and then go back in the reverse direction. Repeat with the second dowel.

PRO TIP: If you have a partner in crime, have them spin the dowel so you can just hold the roll of tape and let it feed itself.

3. Color the exposed ends of the dowels with your Sharpie.

4. You're going to need only four links of handy link chain, so if your piece is longer, use a hacksaw to cut it.

5. Use two pairs of pliers to grab hold of an eyebolt and twist in opposite directions, opening up the circle so one end of the chain can fit inside. Use the pliers again to close the gap. Repeat with the second eyebolt on the other end of the chain.

6. Using a ³⁄₃₂-inch bit, drill a pilot hole in the center of one end of each dowel.

7. Use pliers to screw the eyebolts into the holes until the bolt head rests against the wood.

PRO TIP: For extra strength, put a little bit of wood glue in the holes before screwing in the eyebolts.

FUN FACT: Although nunchucks have been used as deadly weapons by Bruce Lee and characters on TV like the Teenage Mutant Ninja Turtles, they were most likely first created by the Japanese as an Okinawan horse bit.

8. A black handle by itself looks pretty cool, but you can really dress it up by adding a splash of color to the tops and bottoms. Customize any way you like with colored electrical tape!

USING THE NUNCHUCKS

There are plenty of cool sequences to try with nunchucks, and I highly recommend checking out videos of some real masters. If you are a beginner, try holding one handle of the nunchucks and swinging the other back and forth around the front and back of your waist. Then throw it over the same-side shoulder and grab the bottom nunchuck with your opposite arm under the elbow as it swings up. Whip around to the other shoulder and catch the bottom with the opposite arm. Repeat this process enough and it will be become second nature.

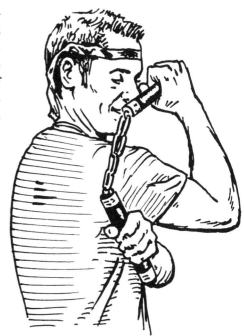

ALSO TRY: If you are nervous about flying pieces of wood, get a piece of pipe insulation foam and a ¾-inch dowel that will fit nicely into the center. If you reinforce those with electrical tape, you can turn your nunchucks into "funchucks," and make them a little safer.

Save the day with these awesome customizable nunchucks. Cowabunga, dude!

Declare miniaturized warfare with a mini airsoft seltzer grenade. The materials for this project are as simple as they can get, but when combined correctly, you'll have an erupting explosive device.

MICRO AIRSOFT GRENADE

04

SAFETY KEY:
+ Safety glasses

SKILL LEVEL:
EASY
INTERMEDIATE
ADVANCED

APPROXIMATE TIME:
15 minutes

WHAT YOU'LL NEED:
+ Alka-Seltzer tablets
+ Plastic film canister
+ Roll of duct tape
+ Airsoft BBs

LET'S BEGIN

MAKING THE GRENADE

> **HELPFUL HINT:** Film canisters are a perfect size for this project and have lids that snugly secure. If you cannot find them, try a comparable plastic container with a snap-on lid.

1. Cover your canister with duct tape. Use scissors to cut out a circle for the top and bottom. This will give your grenade a cool texture, as well as increase its durability.

> **PRO TIP:** Duct tape is the exact width of a film canister, which will make wrapping it up very simple.

2. Drop a single Alka-Seltzer tablet and a handful of around forty to fifty BBs in your film canister. Add warm water to catalyze the reaction, pop the lid on, and give it a little shake.

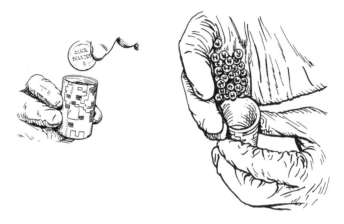

3. Just as with a real grenade, you'll have about five to six seconds to get to a safe place (approximately 15 feet away) to watch the explosion!

SAFETY NOTE: Always wear protective safety glasses when using this grenade.

PRO TIP: If you can get to it in time, you should be able to use the same Alka-Seltzer tablet for another round.

HOW IT WORKS

Alka-Seltzer tablets contain baking soda and citric acid. When the tablets are dissolved in water, the baking soda (or sodium bicarbonate, to be fancy) and citric acid in the tablet react with the hydrogen in the water to create carbon dioxide gas. The carbon dioxide builds pressure in the container until it is substantial enough to burst the container.

KICK IT UP A NOTCH: Turn your grenade into a high-flying rocket by placing it cap side down on the table before it bursts.

Who doesn't like trifling with Alka-Seltzer rockets? After some improvisation with materials, the prototype was finally modified into a mini desktop grenade. The anticipation of the container bursting makes it a forceful and unique device that you'll look forward to experimenting with!

FUN FACT: The stun grenade was designed as a non-lethal incapacitant in the 1960s by the British Special Air Service.

Build up your explosive desktop arsenal with a mini rocket-propelled grenade (RPG). It's so simple, you probably have the materials lying around the house. So, what are you waiting for?

MINI RPG LAUNCHER

05

SAFETY KEY:
+ Explosives + Fire

WARNING:
☠ These mini RPGs are only intended to be shot at paper targets.

SKILL LEVEL:
EASY
INTERMEDIATE
ADVANCED

APPROXIMATE TIME:
20 minutes

WHAT YOU'LL NEED:
+ Straws
+ 1 Larger straw
+ 3 LEGO blocks
  (6 studs each)
+ Barbecue lighter
+ Hot glue gun and glue
+ Pop-Its (small novelty fireworks you may know as "bangers," "poppers," "party snaps," etc.)
+ Duct tape

LET'S BEGIN

MODIFY YOUR STRAW

1. Snap three LEGO blocks together—one at the top and two on either side. Line up your straw so it is even with the top LEGO and cut off the end.

2. On to the fins! Make four cuts along the lines of the straw, two LEGO studs deep. This doesn't need to be super precise—all you need are four flaps that will fold upward and form a cross.

3. Heat-treat them with the barbecue lighter. Ignite your flame and slowly lower the straw toward it from the top until you see the fins start to move downward. Then quickly push your thumb up onto the fins and hold for a few seconds, allowing the heat to re-form the plastic.

4. Spread your Pop-Its on the table and give the sawdust a little blow. Insert a Pop-It into the front of the straw with the paper streamer facing down. Secure into place with a dab of hot glue.

FUN FACT: Pop-Its come encased in sawdust because they are essentially just bits of gravel covered with fulminated mercury, a simple contact explosive.

5. Gather the fins of the RPG back into a straw shape and then slowly push them into the second, larger straw.

6. All you need is a deep breath of air and you're good to go! The best thing about these cartridges is they are reusable. After you snap one off, pull out the leftover hot glue, grab a new Pop-It, add a dab of hot glue, and reload the exact same way.

SPECIAL FEATURES

1. Your straw works great, but needs to look a lot cooler before it becomes a bona fide RPG launcher. Tear off a piece of duct tape an inch or so longer than the straw. Put the strip sticky-side up on a table and gently roll the straw in it until covered. For a professional finished look, add a strip of electrical tape on both ends, and a couple in the middle just for decoration.

2. Make an optical scope by wrapping more electrical tape around a thin, clear straw. Cut the straw at an angle and hot glue it to the end of the launcher.

3. Use a black marker to color both ends of a Popsicle stick black. Cut about half an inch off the tip of the Popsicle stick, and then cut that in half lengthwise. Color any exposed wood, and then glue the makeshift handle so when the launcher is flipped upside down, the handle looks like a shark fin with the scope hanging directly below it. Glue the second piece about halfway down the launcher.

Your upgraded launcher will work exactly the same as before, as it is literally the same straw. Pinch the fins, push the RPG down into the launcher tube until the warhead rests against the tip, grab the handles, place it in your mouth, and give a quick burst of air.

SIDE PROJECT: Put them to the test with your own eerie paper targets like *Minecraft* zombies or creepers!

NEXT LEVEL: If you want to go extreme, find Pop-Its with even more eruptive potential. Some contain ten times as much fulminated mercury. They will load the exact same way but will create a bigger bang and an even more intense explosion!

FUN FACT: A rocket is a chamber enclosing gas under pressure. Through a small opening at one end of the chamber, gas is allowed to escape, causing a thrust that propels the rocket forward. The simplest example is releasing the air from an inflated balloon.

Super shooters! With bottle caps and balloons, you can easily make yourself a powerful pocket slingshot. So grab a marshmallow and start shooting!

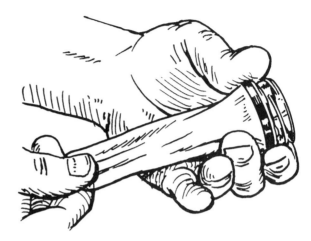

POCKET SLINGSHOT

SAFETY KEY:
+ Adult supervision + Projectiles + Safety shades

WARNING:
☠ Avoid using ammo harder than a marshmallow because as innocent as this looks, it packs quite a punch and it'll leave welts.

SKILL LEVEL:
EASY
INTERMEDIATE
ADVANCED

APPROXIMATE TIME:
20 minutes

WHAT YOU'LL NEED:
+ Wide-mouth juice bottle   + Latex balloons
+ Hacksaw   + Mini marshmallows
+ Sandpaper

06

LET'S BEGIN

BUSTING CAPS

1. Get yourself one of the juice bottles with a really wide mouth. Any kind will work, but Gatorade bottles seem to have some of the biggest mouths.

2. Use a hacksaw or a pair of scissors to cut through the bottle just below the top so it's flush with the bottom of the plastic flange.

PRO TIP: If you don't have a hacksaw, scissors will work, but you run a greater risk of cutting yourself.

3. Use some sandpaper, or the sidewalk, to smooth down the cut side of the plastic flange. Rubbing it around in circles will remove any plastic burrs and get it ready for adding the balloon.

4. Use your fingernails to remove the plastic band around the cap but don't cut or break it. You'll need this again a little later on.

1. Find a balloon, any color you want, and snip the neck off about a quarter inch above the main body curve.

2. Stretch the balloon and pull it around the flange until the mouth lines up with the top edge of the bottle.

PRO TIP: Make sure the balloon is perfectly centered, and smooth out any wrinkles around the threads. If the balloon is out of alignment, your ammo may not shoot straight.

3. Replace the plastic band over the balloon so it's back in position at the bottom of the threads. Then roll the top of the balloon down until it nestles neatly into the groove just above the plastic band.

4. Give your balloon a test tug from the bottom. When you pull the balloon, the ring should hold the top firmly in place.

READY, AIM, FIRE!

1. Your Pocket Slingshot is finished and ready for action. Hold the cap between your thumb and forefinger and load it by dropping a mini marshmallow into the balloon. Then pull back, aim, and release!

2. Practice makes perfect. Focus on pulling the balloon straight back so that upon release, it springs completely through the bottle top to the other side. Pulling at an angle could cause the balloon to snap back against your fingers, and that doesn't feel great.

3. When you're done using your slingshot, you can store your ammunition in it as well. Just load it up, screw the cap back on, and collapse it into a little puck that slides easily into your pocket. It is portable, inconspicuous, and tons of fun.

WARNING:

☠ It's important not to shoot projectiles at people or animals. These things can hurt a lot. So have fun, but be responsible about it.

> **FUN FACT:** Loading your pocket slingshot with a pinch of salt can remove pesky flies from your house without damaging the walls.

Powerful pocket slingshots are cool but can be expensive. What makes this super shooter especially mind-blowing is that it can be made on a budget *and* the good times are never compromised. Persistence and curiosity always pay off!

This magical mixture handles like pizza dough, but the instant it goes still it liquefies and melts into a glowing goo!

MAGIC MUD

SAFETY KEY:
+ Knife

SKILL LEVEL:
EASY
INTERMEDIATE
ADVANCED

APPROXIMATE TIME:
30 minutes

PRO TIP: It's mind-blowing to think this stuff is found inside our potatoes, but if you don't feel like going to the trouble to get it out, cornstarch will work exactly the same.

WHAT YOU'LL NEED:

+ Large bag of potatoes
+ 2 large mixing bowls
+ Large strainer
+ Tonic water

+ Small bowl
+ Spoon
+ Food processor
+ Tall jar with lid

07

LET'S BEGIN

CONCOCTING YOUR MUD

1. Wash the outsides of your potatoes so they are all nice and clean.

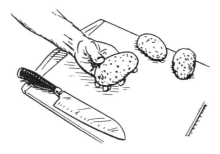

2. Use a food processor (or knife and cutting board) to chop them up into the smallest pieces possible.

3. Slide the potato pieces into a large mixing bowl and add enough hot water to cover completely. Stir for a few minutes. You may notice the water turning red—this is normal.

4. Grab a strainer and another large bowl and separate the potato water from the potato pieces. After about ten minutes a strange white layer will appear at the bottom of the bowl of potato water. Pour out the water and the white goop will stay at the bottom.

5. Mix a bit of fresh water into the white goop to help with the dryness, and pour the mixture into a tall jar with a lid. Screw the lid on tight, shake it all up, and let it sit for another ten minutes.

6. Dump out the water, which will take most of the impurities with it, leaving you with a very clean and magical substance. It looks like milk, but if you try stirring it around, you can see it behaves very strangely. It will look and feel like pizza dough, but pause for just a second and the mixture will collapse into goop that slimes between your fingers.

7. To add an awesome glow, leave your mixture out for two days, and it will turn into a white powder. Add a few spoonfuls to a clean bowl and then mix in a splash of tonic water. In just a few seconds it will become difficult to stir, and with a little patience, everything should combine and behave exactly the way it did before—the difference is that now it's fluorescent. If you turn on some black lights, the Magic Mud takes on a mystical glow.

FUN FACT: Tonic water adds that spooky glow because it contains quinine. If you put a bottle of tonic under a black light, the tonic water fluoresces, and the whole bottle will look completely supernatural.

HOW DID WE COME UP WITH THIS? This project was thought up while attending a sales presentation for pots and pans, where one of the pots, a potato dish, was coated with a strange white residue. Out came the curiosity, and the experimenting began. You never know when your inquiring mind will lead to a magical gooey substance!

These homemade sticks of pampered bliss will instantly turn an ordinary bath into a rich and relaxing spa experience. They are easy to make and smell incredible, and are the perfect way to show a special someone just how much you care. Time to drop the love bomb.

TNT BATH BOMB

TNT BATH BOMB

SKILL LEVEL:

EASY
INTERMEDIATE
ADVANCED

APPROXIMATE TIME:
45 minutes

08

WHAT YOU'LL NEED:

+ ¼ cup baking soda
+ ⅛ cup Epsom salt (or sea salt)
+ ⅛ cup citric acid
+ ⅛ cup cornstarch
+ 1½ teaspoons vegetable oil
+ 12–15 drops red food coloring
+ 1 teaspoon water

+ Measuring cups
+ Bowl with lid
+ Paracord or string
+ PVC tubing,
 ½ inch wide by
 3½ inches tall
+ Essential oil for
 fragrance (optional)

PRO TIP: Vegetable oil works great, but bonus points if you use castor oil instead. And if you don't have PVC tubing, a toilet paper roll will do the trick.

LET'S BEGIN

MIXING YOUR INGREDIENTS

1. Combine the baking soda, Epsom salt, cornstarch, and citric acid in your mixing bowl.

2. Pop the lid on the bowl and shake it as hard as you can for around twenty seconds.

FUN FACT: Citric acid, or sour salt, is made from ground-up citrus fruits, and is the stuff on sour candy that makes your lips pucker up when you eat them. You can find it at health food stores or online.

3. In a smaller container (like a shot glass), combine the oil and water.

4. Now add ½ teaspoon of the essential oil you want for fragrance and relaxation. This is a matter of personal preference. Eucalyptus smells great and makes your skin tingle.

PRO TIP: Try taking your aromas to the next level by mixing fifteen drops of eucalyptus, five drops of peppermint, and twenty drops of lavender.

5. Now add your food coloring. Fifteen drops of red is enough to turn your bath bomb pink without having to worry about it staining your skin or your bathtub. Mix with something like a toothpick.

6. Add your wet mixture to the dry mixture. When combined the concoction will begin to fizz. To minimize this effect, try stirring the powder continuously while slowly dribbling the oil into the mixture. Mix all the liquid into the powder and then put the lid back on and give it another twenty seconds of vigorous shaking.

FORMING YOUR BATH BOMB

1. Cut a seven-inch piece of string or paracord. Tie a simple knot as close to the end as you can get it.

PRO TIP: If you want, you can tie a tiny secret surprise onto the string and have it pop out in the bath once everything dissolves.

2. For the mold, use a piece of PVC tubing. Cut a piece of paper so it is just wider than the tube is tall, then roll it up and push it down into the center of your mold. When you let go, it will expand and fill the inside of the tube completely.

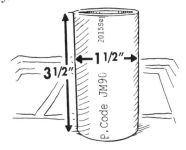

3. Hold the tube upright on a tabletop and carefully spoon the colored compound into the center, one scoop at a time. As it gets near the top you will need to firmly compact the mixture back down to the bottom with something blunt, like the butt of a hammer.

4. Continue filling the tube. When it is almost full, poke a small hole in the center and shove the string halfway down inside.

5. Using the butt of your hammer, ram the mixture slowly and carefully on either side of the string to help secure the rope in place, then press the other edges down. This recipe will give you just enough powder to fill the tube all the way to the top.

6. When it is packed down as hard as you can get it, move the container to the freezer and let it sit for fifteen minutes.

1. Once your bath bomb is done freezing it's time to remove it from the tube. Using the butt of the hammer, apply firm and constant pressure at the base of the bomb. The package should pop out cleanly and in one piece. Remove the paper and you've got a cool little bath bomb ready for use.

2. To make it a touch more professional, add a custom label with a bit of glue at the bottom tab.

TO USE

Simply toss your bath bomb into a tub full of water. The citric acid will react with the baking soda to rapidly release carbon dioxide gas, instantly infusing the bath with color, fragrance, and relaxation. For a moment in time, you or a special someone will be immersed in a blissful utopia of rejuvenation.

ROMANCE, SHMOMANCE: There's no rule that your bath bombs need to be pink or romantic. You can use blue food coloring and a novelty ice cube mold to make any bath detonator, like an exploding **Death Star!**

This TNT Bath Bomb may be a great way to unwind and pamper yourself or someone else, but remember, most important, it's all about the bomb, baby!

Make bouncy, stretchy rubber in an instant! Proto-Putty is a cool, moldable dough that you can press into nearly any shape, and after ten minutes it magically turns into bouncy rubber!

PROTO-PUTTY

09

SAFETY KEY:
+ Well-ventilated area + Gloves required
+ Adult supervision advised

WARNING:
☠ Construction silicone can be harmful if swallowed, inhaled, or absorbed through the skin. Proto-Putty is good for making toys, but not good for making candy molds.

SKILL LEVEL:
EASY
INTERMEDIATE
ADVANCED

PRO TIP: It's extremely important to use silicone #1, so you should double-check that your tube says "100 percent silicone." This is critical to your success.

APPROXIMATE TIME:
10–25 minutes

WHAT YOU'LL NEED:
+ Cornstarch
+ Tube of silicone #1—
   100 percent silicone
   (Important: silicone #1 only)
+ Caulking gun

+ Food coloring
+ Disposable paper bowls
+ Popsicle sticks
+ Disposable rubber gloves

LET'S BEGIN

GET MIXING

1. Lay out two paper bowls. Fill one halfway with cornstarch and set it aside.

2. Now squirt a generous amount of food coloring into the second bowl. The more you use, the richer the color will be.

3. Put your rubber gloves on and carefully squeeze a big blob of silicone directly into the food coloring. Use a Popsicle stick to slowly fold the silicone and food coloring together.

HELPFUL HINT: Squirt out only as much silicone as you think you'll need for your project.

4. Act quickly. General-purpose silicone is activated by moisture, and food coloring begins the curing process, so you'll only have around five to ten minutes of work time before the mixture starts hardening. The food coloring is fully mixed when it's blended into a uniform color.

PRO TIP: Silicone #1 releases acetic acid vapors while curing and smells so strongly of vinegar that it will sting your nostrils. The food coloring will also stain your clothes or the area around you, so consider mixing this outside, or at least in a well-ventilated area.

5. Scoop the blob of colored silicone into your bowl of cornstarch and generously coat the outsides with the powder to prevent the silicone from sticking to anything.

6. Flip it over a few times and knead it into different shapes, adding more cornstarch each time it's about to start sticking to your gloves.

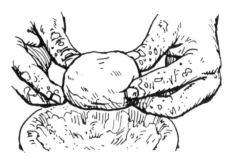

PRO TIP: The mixture is extremely sticky and difficult to work with at first, but give it a couple of minutes and plenty of cornstarch, and it'll start to firm up into dough.

7. Repeat the kneading process ten to fifteen times. After around two minutes, the color of your silicone should brighten back up and it should start to feel like Play-Doh.

8. You can now mold it, squish it, or press it into whatever shape you want (don't forget the time constraint to make your shape).

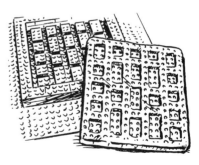

HELPFUL HINT: Try rolling little balls to make your own high-flying bouncy balls. Or make another batch with neon food coloring to make some cool colors, or to make it glow in the dark.

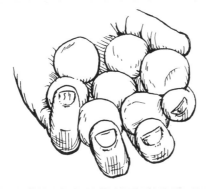

PRO TIP: Proto-Putty works great for impression casting, but don't use it for making custom candy molds. Food-grade silicone is available to purchase online instead if you're tempted to make molds for chocolate or other edible treats.

FUN FACT: Proto-Putty is bouncy, stretchy, and fairly durable, especially the thicker you make it. It'll still cure without adding any cornstarch at all, but it's super sticky and hard to work with. Silicone caulk by itself takes weeks to firm up. The trick is simply adding a bit of food coloring!

GET CREATIVE: You can make anything you can think of with Proto-Putty. And now that you know how to make it, you should be able to fix, invent, or play your way into a whole new world of creativity. It's all up to you!

If you're on a hike or caught in a survival situation, it is helpful to have containers that are lightweight and weather resistant. In this project let's turn a couple of empty soda bottles into compact, waterproof canisters.

SODA CAP CONTAINER

10

SAFETY KEY:
+ Sharp objects

SKILL LEVEL:
EASY
INTERMEDIATE
ADVANCED

APPROXIMATE TIME:
20 minutes

WHAT YOU'LL NEED:
+ Hotel key
+ 2 plastic soda bottles
+ 150-grit sandpaper
+ Two-part epoxy
  (a two-component adhesive
  that comes in one dispenser)

+ Bench vise
+ Hacksaw

LET'S BEGIN

BUILDING YOUR SODA CAP CONTAINER

1. Clamp a soda bottle upside down in your bench vise.
Using the bottom side of the flange as a guide, cut off the
bottle at the neck as cleanly as possible with a hacksaw.
Repeat with the second bottle.

PRO TIP: Use two bottles with different-colored caps so you can easily
differentiate the compartments.

2. Sand down the rough plastic edges so the lids will fit
together perfectly. Sand the hotel key as well—the grit
will allow it to bind better with the epoxy.

3. Carefully trace one of the caps with a marker onto the hotel key. Cut out the plastic circle. This will be the divider between the two compartments.

4. To make it as strong and lightweight as possible, bond using two-part epoxy. Blend the epoxy thoroughly and quickly and apply a liberal amount to the base of each cap, then carefully set the round divider on top, and press it into place. In about two hours, your container will be firm enough to handle, and a few hours after that it should be fully ready to use.

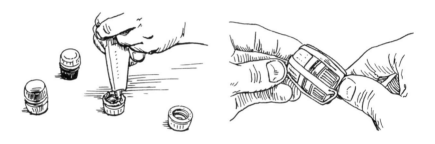

PRO TIP: For an even stronger bond, clamp both caps using a bench vise, set of bar clamps, or rubber bands so they set together under pressure.

ALSO TRY: An even simpler container can be made with two caps and some hot glue. It won't be as strong, but you can make it in under three minutes from start to finish. This version is still super lightweight, water resistant, and works well for storing things like pills or small candies.

SODA CAP CONTAINER ▮

PUT IT TO USE! The container weighs a mere half ounce, unscrews from both ends, and is completely waterproof. You can utilize it to store compact items like materials to start a fire or any other type of emergency kit! Whether you're simply organizing your backpack or find yourself in an apocalyptic situation, these little capsules make life a whole lot easier.

FUN FACT: People around the world buy one million plastic bottles every minute. These bottles take four hundred years to decompose. So, instead of throwing that bottle away, you can recycle it to make a cool container that's waterproof, airtight, and will last for four hundred years!

Mash them! Juggle them! Customize them! Just grab a bag of flour and party balloons and create a colorful batch of superhero stress balls at home for practically nothing!

NINJA STRESS BALLS

11

SKILL LEVEL:
EASY
INTERMEDIATE
ADVANCED

APPROXIMATE TIME:
15 minutes

WHAT YOU'LL NEED:
+ 2 wide-mouth juice bottles
+ Bag of latex party balloons
+ Bag of flour

LET'S BEGIN

ASSEMBLING YOUR NINJA STRESS BALLS

1. Cut a plastic water bottle in half and then turn it upside down, forming a makeshift funnel.

2. Use the funnel to transfer ¾ cup of flour into an empty wide-mouth juice bottle.

3. Blow up a balloon, twist the end a couple of times, and stretch the spout over the mouth of the juice bottle. When attached, let the balloon go, turn the bottle upside down, and squeeze all the flour into the balloon.

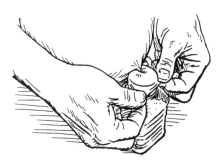

4. Pinch the neck of the balloon and pull it off the bottle to let the air out. Keeping the neck pinched, massage the base to compress the flour and move any trapped gases to the top.

5. Clean the outside of the balloon with a damp cloth, then carefully use a pair of scissors to cut off the neck. While you still have the scissors out, cut the ends off a few more balloons of the same color.

6. Use all your fingers to stretch one of these balloons open and place over the center of the flour ball. Wrap it around to the underside to seal in the flour. Pull the edges to smooth out any ripples. Repeat this step with two or three more balloons to increase its durability.

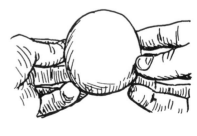

PRO TIP: Try adding a few dabs of superglue on the underside of the lip to make your balloon extra secure.

---

7. Finally, add a "ninja mask." Cut U shapes around the edges of a black balloon, then fold it in half so you can cut a U shape out of the middle. Putting the mask on the Ninja Ball is as simple as turning the band inside out and wrapping it around the outside of the ball.

---

8. Try matching one of the holes of the mask with the seam on the ball, and the seam will virtually disappear. This will give it a professional touch, and most people will have no idea how you made it.

ALSO TRY: Play around with different color combinations and firmness levels. See what happens when you fill a Ninja Ball with sand or sawdust instead of flour. Try four solid green balls with blue, purple, red, and orange masks to get Ninja Turtle eggs, or think up any other superhero color combos. The options are limitless. Why pay $10 for a pack of three when you can customize your own for ridiculously cheap? Plus, these come out better than the ones you see in stores! They're soft, durable, and ultrafun to play around with.

WARNING:

☠ So awesome and easy, you may become addicted.

> **FUN FACT:** Did you know the first stress ball was actually a walnut? During the Han dynasty (206 BC), warriors would squeeze walnuts because they believed it would enhance their fighting skills in battle.

Whether you're launching rockets, sending action figures into battle, or throwing Sky Ballz in the park, this super simple parachute will give you a soft landing every time!

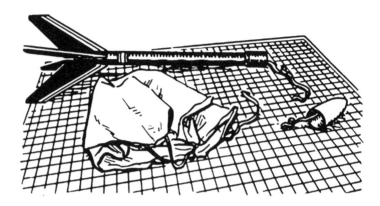

TABLECLOTH PARACHUTE

SKILL LEVEL:
> EASY

INTERMEDIATE

ADVANCED

APPROXIMATE TIME:

15 minutes

WHAT YOU'LL NEED:

+ Rectangular plastic tablecloth

+ #3 crochet thread

+ #7 barrel swivels

+ X-Acto knife or scissors

+ Scotch tape

12

LET'S BEGIN

TRIMMING AND SHAPING

1. Remove the tablecloth from the packaging. The material will be smooth and folded on one end, while the other side has all the straight cuts. Keep in mind which is which. Without unfolding the tablecloth, measure it thirteen inches from the folded end to the loose end. Cut the excess length. Double-check that the side you're cutting is the one with the loose ends. For the cleanest cut, use an X-Acto knife and something like a wooden board as a straightedge.

2. Unfold the tablecloth lengthwise while keeping it the same width. Cut off exactly twenty-six inches from one of the ends, leaving you with a rectangle that is thirteen by twenty-six inches.

3. Now it's time to use a little origami. Fold the rectangle in half lengthwise to make a square. Make sure the bottom-right corner has the four loose ends. Hold the corner in place with your index finger and then fold the square point to point, forming an isosceles triangle. Tilt the triangle to have a straight right edge, making sure the loose ends are on that side. Measure five inches from the bottom, double-check that the loose ends are on the right side, and then go ahead and cut the bottom piece off.

PRO TIP: You could always be resourceful and use these scraps for making micro parachutes.

4. When you unfold the triangle, you'll see a little square chunk missing from the bottom-right corner. When you unfold it two more times, you should end up with a cross. You've just made two parachutes. Quickly make six more by repeating this process with the tablecloth leftovers.

ASSEMBLY

1. Tear off a two-inch piece of Scotch tape and secure the upper-left section of it to one corner of the parachute. Pull over the adjacent corner and line up the two sides— the edges should match perfectly at the bottom. Use the remaining upper-right corner of the tape to connect the two sides together. Now flip the canopy over and fold the other half of the tape over to stick it flat to the inside of the parachute as well.

2. Do the same thing with all four corners. Your canopy is starting to take shape.

RIGGING THE PARACHUTE CORDING

1. Cut two strings that are each thirty-three inches long.

PRO TIP: I recommend using #3 crochet thread, but most other strings will work as well. It's also a good idea to mark both ends of the string one and a half inches from the bottom because these markings will make it much easier when it comes time to attach them to the canopy.

2. To attach the lines to the parachute, press a small piece of tape onto the strings so the bottom of the tape lines up with the one-and-a-half-inch mark. Then attach the string to the inside of the parachute by pressing it flat to the plastic cloth, making sure the mark lines up at the very bottom of the canopy. The excess tape needs to be folded over and secured to the canopy the same way. Use another two-inch piece of tape and flatten the string down over the top of the canopy. The other end of the string gets attached the exact same way to the corner right beside it. Repeat these steps to attach the second string to the remaining corners.

3. Bring two of the attachment points, connected by the same piece of string, together so the marks on either end of the string match up, then pull the parachute strings completely straight. Without letting go of the string, carefully tie a knot near the end, making sure to leave a small loop at the bottom. Repeat with the other string.

4. Line the two loops beside each other and thread them with some elastic cording. Secure with a couple of square knots. Pull really hard to make sure this connection never comes undone.

5. Now trim the excess, making sure to leave enough room on one end to tie one of the swivel hooks the same way. Your parachute is now ready to attach to a Sky Ball, action figure, or rocket, and bring them safely down to earth.

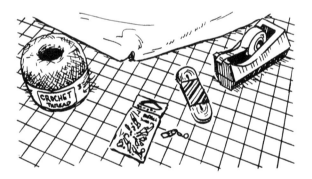

PRO TIP: You can attach barrel swivels to *both* ends of the elastic cording, and make them any length you need. This way, if your lines get tangled to the point of frustration, they will be easy to unclip and fix within seconds.

PACKING THE PARACHUTE

1. Hold the parachute at the center and run your hand down the parachute, squeezing it all together, and roll it up. Secure the canopy by wrapping the strings around the outside, which will hold it together until it deploys.

These cheap and incredibly effective parachutes add a whole new dimension to the fun you can have outdoors. It's never a dull day at the park when you're equipped with an endless supply of canopies for all your aerial activities!

> **FUN FACT:** In 1783, Louis-Sébastien Lenormand was so confident in his new invention of the parachute that he introduced it to the world by jumping from a tower.

Throw one of these Sky Ballz high into the air and a little parachute opens up, bringing them safely back to the ground. It's time to utilize your devices from other projects for the ultimate elevated experience.

SKY BALLZ

SKILL LEVEL:
EASY
INTERMEDIATE
ADVANCED

APPROXIMATE TIME:
15 minutes

WHAT YOU'LL NEED:
+ 1 Ninja Stress Ball (from chapter 11)
+ 1 Tablecloth Parachute (from chapter 12)
+ 1 latex balloon

OPTIONAL:
+ Swivel hooks
+ String

LET'S BEGIN

MODIFY YOUR NINJA STRESS BALL

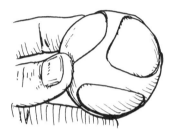

1. Remove the mask from one of your Ninja Stress Balls.

2. Open the mouth of your latex balloon wide enough that you can fit the entire Ninja Stress Ball inside. Then tie it off at the top.

3. Fit the ninja mask back over the ball, leaving the tied-off portion of the balloon exposed at the top so you can tie a string to it.

1. Look for the two loops at the end of your parachute cord and carefully thread a piece of string through both of them and tie it off.

2. Tie the other end of the string to the Ninja Stress Ball, making sure the string is attached underneath the knot. Cut off the excess string and your Sky Ball is finished.

TIME TO PLAY

1. To pack the parachute, simply hold it at the top so the strings hang down straight. Squeeze all of the air out of the material, fold it in half, then start rolling the parachute up into a ball toward the strings. When you get to the strings, carefully wrap those around the parachute as well. This will keep the bundle together, but will also unravel fairly easily when you throw it.

2. The next part is effortless! Simply throw the bundle high into the air and watch your parachute open up and float gently back to the ground.

HELPFUL HINT: The tighter you pack your chute, the longer it'll take for it to open.

PRO TIP: If you've made the Skyblaster Slingshot from chapter 30, use it to see how high you can shoot your Sky Ballz. But make sure to pack and wrap them extra tight so they don't open too early.

NEXT LEVEL: If the lines start twisting after a bad throw, it's not hard to unravel them. But if you want to go one step further, just add a barrel swivel to your balloons. These clips make it really easy to unhook and untangle the lines in a hurry. You can find them in the fishing aisle of your local supercenter.

Playing with these Sky Ballz and the Skyblaster Slingshot will attract a lot of attention from onlookers, and it might just make you the coolest one at the park.

The world's simplest semiautomatic rubber band handgun just became the most effortless weapon to add to your desktop arsenal!

RUBBER BAND HANDGUN

14

SKILL LEVEL:
  EASY
  INTERMEDIATE
  ADVANCED

APPROXIMATE TIME:
  5 minutes

WHAT YOU'LL NEED:
  + Just some rubber bands and your human hand

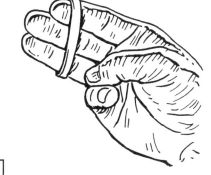

LET'S BEGIN

LOCKING AND LOADING

1. Make your hand into a simple gun shape with your thumb and index finger extended and your three remaining fingers curled into your palm.

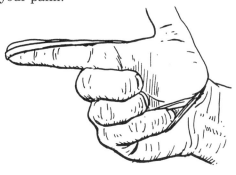

2. Hold one end of a rubber band with your pinky finger, then stretch it back and around your thumb while positioning the other end on the tip of your pointer finger.

3. Use your pointer finger to aim at the target, then simply open your pinky finger to fire the shot off.

ALSO TRY: With a slight variation in your shooting technique, you can easily switch from single shot to semiautomatic to shotgun blast. Start by loading a rubber band the same way you did before, then put two more rounds in the chamber with a second and third band on your fourth and middle fingers. Fire in quick succession for semiautomatic, or all at once. With this setup you can fire at multiple targets without having to stop and reload after each shot.

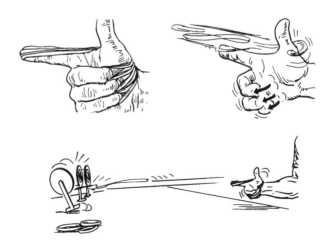

HELPFUL HINT: Make your targets out of construction paper and paint stirrers if you already have them lying around, but anything that's not breakable or alive will work. Try starting off a domino chain or constructing zombie targets! The world is your rubber band shooting range. Besides, you already have the most important tool on hand.

FUN FACT: Rubber bands were patented in England on March 17, 1845. The US Postal Service is said to be one of the biggest consumers of rubber bands on earth, using them to sort and group mail.

This sticky green goo is so easy to make even a three-year-old can do it! All you need are some common household items and you've got yourself an iridescent ooze that's extremely fun to play with.

NINJA TURTLE OOZE

15

SAFETY KEY:
+ Nonedible: Ninja Turtle Ooze is nontoxic and doesn't really taste bad, but it's probably better if you kept it in your hands and out of your mouth.

SKILL LEVEL:
EASY
INTERMEDIATE
ADVANCED

APPROXIMATE TIME:
15 minutes

WHAT YOU'LL NEED:
+ One 5-oz. bottle Elmer's nontoxic school glue
+ Food coloring (green and yellow)
+ Borax detergent booster

+ Measuring cup
+ Measuring spoons
+ Mixing bowl
+ Two 2-inch ABS adapters

LET'S BEGIN

MAKING YOUR MIXTURE

1. Combine 1 cup of water and ½ teaspoon of borax. Stir well.

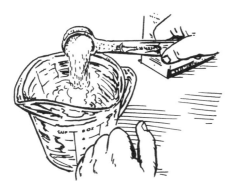

2. In a mixing bowl, mix ½ cup of water and the entire 5-oz. bottle of Elmer's school glue. Add two drops of green food coloring and five drops of yellow food coloring.

3. Combine the two mixtures and watch gelatinous ooze begin to form before your eyes.

4. Transfer your slime to a different bowl. You've now got a mound of semi-transparent ooze that looks like an alien experiment and is really fun to play with.

STORAGE: You can keep your Ninja Turtle Ooze in any airtight plastic container. Play-Doh canisters work great, or you could put it back in the cleaned-out glue bottle.

MYSTERIOUSLY SLIMY! How do you think your neighbors might react if they happened upon something that looked like this? Would they call the police? Or fight the villain themselves? Be ready to find out.

If you were to find yourself in an emergency or a zombie situation, finding clean drinking water would be your first priority. Here's an easy, effective way to build a water purifier in an apocalyptic pinch!

EMERGENCY WATER FILTER

16

SKILL LEVEL:
EASY
INTERMEDIATE
ADVANCED

APPROXIMATE TIME:
90 minutes

WHAT YOU'LL NEED:
+ 3 empty water bottles
+ 3 pie tins
+ Strainer (optional)
+ Charcoal
+ Brick (optional)
+ Sand
+ Paper towel

LET'S BEGIN

SETUP

1. Separate your sand by pouring it into a container with some water. Mix together; the larger rocks will sink to the bottom and the sand and clay will stay on top. You could also use a strainer. Place the strained sand in one of the tin pans.

2. Use something heavy, like a brick, to crush your charcoal into a fine powder. The finer your charcoal, the better the filter will work.

ASSEMBLING THE FILTER

1. Cut the bottom off a clear water bottle. Press a piece of paper towel into the top, near the cap opening, to catch everything. You can also use a bit of your shirt, a sock, or any kind of cloth that can hold the charcoal in place.

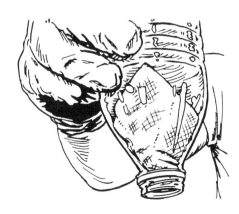

2. Begin adding the filter materials, from fine to coarse: a few inches of charcoal powder, as much of the fine sand as you can pack in, and finally a nice top layer of small pebbles to keep the water from splashing out.

3. To go one step further, cut the bottom of the bottle into a square to make a baffle, which is a wall or screen, to stick on top of the pebbles. This will help to control the water flow and reduce erosion.

ALWAYS A WAY TO UPGRADE:

This is a very basic-style filter and there are plenty of variations and small adjustments to make it even more efficient. For example:

+ Layer the charcoal and the sand in an alternating pattern, as many layers as you want.

+ Cut the water bottle in half, and use the top part of it to form the filter and the bottom to form a cup.

USING THE FILTER

1. Collect some scummy pond water to use as a test subject.

2. The first tool in your filtration arsenal is gravity. Let your water sit for a few hours, allowing most of the larger organic material to sink to the bottom. Transfer the clear liquid to another container, leaving the sediment behind.

3. Pour the water into the filter. It may take about forty-five minutes for the water to work its way through.

4. After the water passes through the filter, it should look clear and refreshing. However, **don't drink it until you sterilize it**. This can be done for four minutes in a microwave, but there are dozens of other methods like water purification tablets or even setting your water out in the sunlight for twenty-four to forty-eight hours.

PRO TIP: Once you're an expert, you could collect the scummiest water from a pond, and after filtering and purifying it, drink a whole glass and live to tell the tale. You're now able to turn contaminated water into a hydrating liquid that might just save your life.

FUN FACT: The human body is 60 percent water, which means it's the major component of most body parts like your brain, heart, skin, and muscles. Even your bones are 30 percent water!

If you ever find yourself in a
survival situation on the beach,
all you need to do is grab a coconut!

COCONUT ROPE

SKILL LEVEL:
> EASY

INTERMEDIATE

ADVANCED

APPROXIMATE TIME:
20 minutes

WHAT YOU'LL NEED:

+ Coconut husk

17

LET'S BEGIN

SETUP

1. Taking a dry coconut, you'll notice the brown husk fluff inside. Using your fingertips, pull off bits of the fluff.

2. Then make it so your fine, fluffy fibers are in a ball.

PRO TIP: You want to get the finest fibers possible. Discard any fibers that are too hard, keeping as much of the fine, fluffy fibers as you can.

MAKING THE BASE OF THE ROPE

1. Using your finger and thumb, begin twisting in one direction from your ball of fibers and give a little tug.

2. As you tug and twist, like magic, the ball of fibers should continue to pull into a string until there is nothing left of the ball. Now you have a small piece of cordage.

STRENGTHEN YOUR CORDAGE

1. The cordage you pulled from your ball of husk fibers is not very strong. To make a stronger piece of rope you will want to do a reverse twist. First you will take your cordage and bend it in half at the middle.

2. Take the left side and twist it in the same direction you were twisting before, which should put some tension on the cord.

3. Next, place that over the other side of the string and pinch it with your finger and thumb.

4. Repeat this with the other side, which now should be on the left. Continue this process until you are out of cordage. The reverse twist locks the two sides together so it cannot come undone.

LENGTHEN YOUR ROPE

1. Now you have a rather short piece of rope. To lengthen your rope you will want to graft the ends to another piece of cordage.

2. You can graft some already made cordage to the end of your rope by rubbing the fibers at the ends of the two pieces together and giving it a little twist.

3. By doing this you can make your rope as long as you want, and you will barely notice where the two ends were grafted together once you reverse-twist it together.

AN EVEN STRONGER ROPE

1. The single reverse twist made stronger rope out of the single cordage, but to make even stronger rope you can repeat the reverse twist again, this time twisting counterclockwise.

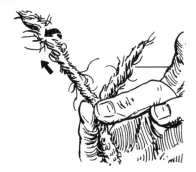

PRO TIP: Pull it to work out some of the rope's tension and repeat the process for even more strength.

2. This time twist the left side counterclockwise and go underneath. It looks like an actual rope, but it was simply woven with your hands!

Who even knew that this tropical fruit could double as a utility or survival device? Rope is extremely versatile. Make it any length or thickness and experiment with just how much weight it can hold or drag. From now on, your friends would definitely pick you if they had to be stranded on an island with someone!

FUN FACT: The coconut has many uses. The natives of the Kiribati Islands used it to create armor, the Japanese launched the coconut as a grenade in World War II, and the coconut husk, when burned, has been found to repel mosquitos.

Whether you want bling, swagger, or just a cool new way to decorate, here is how you can make some cool things with soda tabs. In this project we are upcycling these marvels of engineering into lightweight aluminum chains.

SODA CAN TAB CHAINS

18

SAFETY KEY:
+ Sharp objects

SKILL LEVEL:
EASY
INTERMEDIATE
ADVANCED

APPROXIMATE TIME:
10 minutes

WHAT YOU'LL NEED:
+ Soda can tabs (plain silver or fun colors—you decide!)
+ Scissors

LET'S BEGIN

SETUP

1. For this project, you can go to a local recycling depot and pick up the tabs for free, or you can just collect them yourself at home.

2. To start, you want to break off the rivet mouths on each of the tabs you want to use.

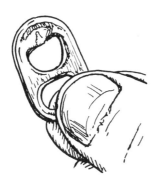

3. Then stack five of them together so they are all in line; this will serve as the first link in your chain.

PRO TIP: You can face the smooth, shiny ends outward on both ends to keep from cutting yourself.

1. Taking a pair of scissors, carefully cut a slit through the thinner end of one tab; this allows you to pry it open, creating a small gap.

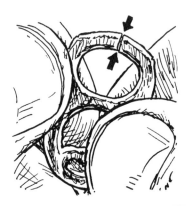

2. Next, all you have to do is push the hook through the thicker end of your chain link and bend the metal back into place.

3. Repeat this process until you have five tabs secured in place.

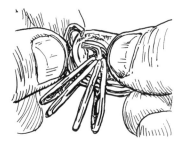

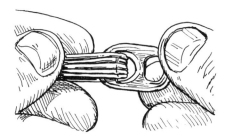

4. Keep repeating this process five tabs at a time until the chain is the length of your liking. It's really that easy!

Once you get the hang of it, there's so much to do with the idea! For example, you could make a custom key chain, a wallet chain, a metallic wristband, a belt, or even a padlock or picture frame! Make them as gifts or keep 'em all to yourself.

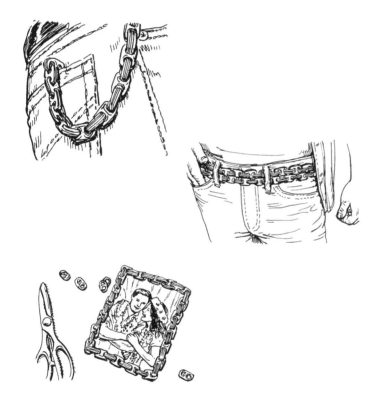

FUN FACT: Soda can tabs are made from one of the most durable metals in the world—aluminum. Aluminum is so sturdy that it has been used to make shark cages, the US Army's Humvee, and the Orion spacecraft.

Is there free energy hiding in your spare change? In this project, you'll learn how to turn a handful of pennies into batteries that could power some of your small electronic devices!

THREE-PENNY BATTERY

19

SAFETY KEY:
+ Electricity

SKILL LEVEL:
EASY
INTERMEDIATE
ADVANCED

APPROXIMATE TIME:
40 minutes

WHAT YOU'LL NEED:
+ Pennies
+ Cardboard
+ Distilled white vinegar
+ Aluminum foil
+ Zinc washers (optional)

LET'S BEGIN

SETUP

1. You will need ten pennies that were produced after 1982 (pennies produced after this year are nearly 98 percent zinc).

2. Next cut some thin cardboard into ten circles the size of the penny and soak them in vinegar.

ADDING THE ZINC

You can expose the zinc in several ways:

+ Take 100-grit sandpaper and sand one side of each penny, exposing the silver zinc. This may take a decent amount of time and energy.

  Or take double-sided tape, stick the pennies to one side, and use an electric sander to sand down the other side. This may cause the tape's adhesive to cover the pennies. No problem! You can use adhesive remover to fix this.

NEED A BREAK? Rather than sanding the pennies you can buy zinc washers for a similar effect.

CREATING THE BATTERY

1. Start with a piece of aluminum as your base.

2. Stack the penny on the aluminum, copper-side down.

3. Add a vinegar-soaked cardboard circle on top.

4. Repeat this until you have used all your pennies and cardboard. (Note: Feel free to use fewer than ten pennies if you need to.)

THE POWER OF A PENNY: If you were to test out the voltage as you go along, you'd find that one penny will generally emit half a volt! With the whole stack of ten, the electrical voltage will jump to about six volts. This is more than enough voltage to drive an LED! You could even light two LEDs at once.

ALSO TRY: If you decided to use the zinc washers, create your battery stack as follows: zinc, cardboard, penny. The penny on the top is the positive side and the washer on the bottom is the negative side.

HOW LONG WILL IT LAST? To test the battery life of your penny stack, use some electrical tape to hold everything in place (with the LED light still connected). Make sure the cardboard edges aren't touching each other and that it is sealed airtight. See how long it takes for your light to burn out.

A PENNY-POWERED CALCULATOR?

1. Use a cheap calculator and remove its battery and the negative and positive leads out of the casing.

2. Make your stacks of battery pennies using either the aluminum or zinc method detailed above.

3. Wrap the stacks with electrical tape and add wires to the stacks and terminals (the positive and negative leads in the casing).

4. Turn on your calculator and test out some functions. If 2 + 2 = 4, we are in business!

There's an idea that hopefully made some cents! If you ever find yourself without batteries, it's worth a shot to use this creative alternative.

In a pickle? Need a fire? Hope you brought a sandwich bag. Take out your pickle and get ready for flames and smoke! Perfect if you're lost on a hike, out of matches at a campsite, or in an emergency survival situation.

SANDWICH BAG FIRE STARTER

20

SAFETY KEY:
+ Fire

SKILL LEVEL:
EASY
INTERMEDIATE
ADVANCED

APPROXIMATE TIME:
25 minutes

WHAT YOU'LL NEED:
+ A sandwich bag
+ Bark
+ Rocks
+ Twigs
+ Dead branches
+ Dry grass

LET'S BEGIN

GETTING STARTED

1. First you'll want to find some tinder to ignite your fire.

PRO TIP: The tinder that you use is critical. It must be extremely dry. Bark is always a good option!

2. Next grab some rocks and start crushing up your tinder into a fine powder.

HELPFUL HINT: The finer you can get your powder, the easier it will light from the rays of the sun.

3. After a few minutes, your powder will reduce to a sawdust-like substance. The darker the powder, the better, because it will absorb more heat!

4. Grab a flat piece of bark for a base plate and transfer your pile of powder onto it.

ALWAYS HAVE A BACKUP: A backup bark plate with powder will come in handy once your embers and coals start burning. You'll want to be able to sprinkle more dust on the fire to keep it burning.

5. Now you're ready to gather some different types of tinder.

SCAVENGING THE FOREST FOR TINDER

Look around and see what Mother Nature has to offer to help you kindle your fire!

+ **Thin twigs:** Usually twigs cover the ground. These are what you will transfer the embers to once you get the flame going.

+ **Dead tree branches:** Break some off from a dry tree and these will also enhance your fire. You can get a full handful in a matter of seconds!

+ **Dry grass:** Search for some dead yellow grass. Yank this out of the ground and try to get enough of it to fold it over and build a nest. The twigs and branches will nestle inside of it.

> **PRO TIP:** Once your fire starts to catch, the smoke and flame will transfer from thinnest tinder to thickest: dry grass, twigs, and then branches.

SANDWICH BAG MAGNIFYING GLASS

No turkey on white for this sandwich bag—it's about to become a tool of combustion!

1. Make sure your bag is empty and then fill it halfway with water. You can use water from a stream or a water bottle.

EXTREME SITUATIONS: Can't find a water source? Ur-ine luck! In an emergency, pee will work just the same.

---

2. Tilt your bag to the side to form a diamond shape with one of your points facing down.

---

3. Then grab the top area of the bag and twist it to trap as much of the water inside as possible. The more you twist your baggie, the more it starts to bulge out until it resembles a liquid sphere.

PRO TIP: There is a fine line between twisting and creating a sphere shape. You want to make it as sphere-like as possible before it bursts!

INVOKE THE POWER OF THE SUN

1. Now that you have your tinder laid out and a cool liquid lens, all you have to do is be patient and let the sun work its magic.

2. Use your baggie bulb as a magnifying glass by hovering it over your tinder. Be careful not to drip water on the powder and bark.

It should start smoking almost immediately!

3. As the white coals start getting exposed, sprinkle some fresh new tinder on top.

4. Repeat this process until you get a good amount of smoke. Keep supplying it with fresh new fuel as it smolders and heats up. Throw together your tinder pile in the meantime.

5. Press the tinder bundle to your smoking bark and carefully turn it over.

6. Gently wrap the bundle around the bark to contain as much heat as possible.

PRO TIP: Thick smoke is a good indicator that it is ready to move on to the next phase. Blow into it gently to accelerate the ignition.

7. Once you get your first flame, add sticks as quickly as possible! If it starts to go out, all you have to do is blow on it a bit more.

There are so many random things to build fires with in survival situations—bottles, lightbulbs, Saran wrap, even your pee! Any time you can add a new item to your fire-starter arsenal, you get closer to becoming a true survivalist. Once you become an expert in the art of fire building, you'll be able to utilize almost any available resource if the situation calls for it.

FUN FACT: How was fire invented? When oxygen reacts with a fuel source, a natural chemical reaction occurs and fire is created. So it really wasn't *technically* invented. Fire from lightning and other natural sources must have amazed the first beholders of fire. However, the question of who was actually able to first create fire from scratch using tools and resources is still a mystery to scientists. Sometimes our questions have no definitive answer. We just have to continue to explore.

We've shown you survival mechanisms before, but you've never seen one quite like this! All it requires is a gum wrapper and an AA battery and you'll be able to start a fire in any enduring situation that life throws at you. Of course you can always get a fire started just for fun as well.

GUM WRAPPER FIRE STARTER

21

SAFETY KEY:
+ Fire

SKILL LEVEL:
EASY
INTERMEDIATE
ADVANCED

APPROXIMATE TIME:
90 minutes

WHAT YOU'LL NEED:
+ Gum wrapper
  (one that has a metallic side)
+ AA battery
+ Scissors

LET'S BEGIN

SETUP

1. The trick to get this to work lies in the gum wrapper itself. What you want to find is a wrapper that has a shiny metallic exterior, but a textured fibrous interior.

PRO TIP: Grab a few different brands of gum to see which wrapper gives you the best results.

2. Make sure you have an AA battery.

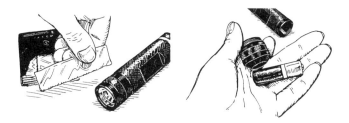

HOW IT WORKS

1. Now, many know that a battery has a positive and a negative end. When those ends are connected to something conductive, electricity begins to flow. That conductive material is going to be the metallic side of your gum wrapper.

THINK ABOUT IT! If we think of electricity like water, then the voltage coming out of the battery is like the pressure of the water and the conductor is like the tube or the piping that takes the water where we need it to go.

2. If you leave the wrapper as is touching both ends, the battery is only going to get super hot and burn your fingertips.

So, you will want to make the wrapper thinner, so that the electrical pressure builds up to the point where it can generate enough heat to start a fire.

DON'T GIVE UP! Finding the right length can be difficult. Too thin and it burns right in half; and too thick and you'll burn your fingers. It's a balancing act.

> **PRO TIP:** What works best is to take your gum wrapper and cut it into thirds. This way you have three different ways to ignite.

STARTING YOUR FIRE

1. Cut your wrapper in thirds, then trim the sides to form an hourglass shape, so the part in the middle is about the same thickness as the blade on your scissors.

2. To get your igniter to work, take your fourth finger and place it underneath your wrapper so the shiny side faces upward.

3. Then take your AA battery and press it down onto your finger, holding the wrapper between the battery and your finger.

4. Then use your thumb and forefinger to hold the battery straight up.

5. This last part happens very quickly. Take the top end of your foil and place it at the top end of the battery. Move your fingers. Now use your thumb to secure the battery vertically, and the wrapper should ignite almost immediately.

PRO TIP: If you hold your battery horizontally, most of the heat will dissipate upward, leaving you with smoke but not ignition.

ALSO TRY: If you are burning your fingers, you can use pieces of the gum or the box the gum came in to protect them.

So what are you waiting for? Plan your own survival fire-starting adventure! Head to the desert with some gum and batteries (maybe some water and fries, too . . . let's not get too risky) and see what you can ignite. Find some dry grass and sticks to light up with your fire starter, and you'll be a survival expert in no time.

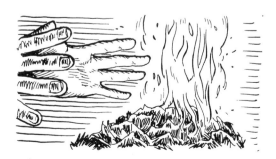

How cool would it be to pour sand into water without it getting wet?! With multiple colors, you can create a host of fun designs and sculptures under water.

MAGIC SAND

22

PRO TIP: You can use a silicone-based spray as well, but I've found that NeverWet spray works best.

SAFETY KEY:
+ Fumes from the spray + Flammable

SKILL LEVEL:
EASY
INTERMEDIATE
ADVANCED

APPROXIMATE TIME:
48 hours (two nights of overnight drying needed)

WHAT YOU'LL NEED:
+ Latex (or rubber) gloves—really any gloves that protect your hands, but also allow you to easily break up the sand without it sticking to the gloves
+ Bag of quartz sand
+ Large tin or aluminum baking sheet
+ NeverWet Multi-Surface Liquid Repelling Treatment by Rust-Oleum (two spray cans included; one is the basecoat and one is the topcoat)

LET'S BEGIN

PREP THE SAND

1. Scatter a few small handfuls of sand on a baking sheet. Make sure you don't put too much, or the spray won't be able to saturate all the sand.

WATERPROOF THE SAND

1. For this step, you'll want to go outside to avoid inhaling the fumes from the can. Once outside, spray the sand that has been distributed flatly across the pan. Make sure you spray evenly across the entire surface of the sand.

☠ **WARNING:**
Because the fumes from the spray could potentially be harmful, make sure you spray down the sand outside. Let the sand air-dry, as the waterproof spray is flammable and you do not want to try to speed up the drying process by adding any heat. Be sure to wear gloves as well since you'll be breaking up the sand with your hands after it's been waterproofed.

2. Let it dry outside overnight.

3. Once it has dried, use your hands to break up any clumps that have formed so the sand is made up of fine particles. Then stir up the sand and spread it evenly across the tin sheet so that you are re-spraying parts that you did not reach in the first go-round. You will want to do this a few times to get a couple of layers of coating on the sand.

4. Let the sand dry overnight again.

TEST IT!

1. Now that your sand is dry, break up any clumps so it returns to a fine powder. You can test it out by putting a small bit of sand on a spoon and dunking it in a bowl of water, or you can dump all of it in the water to see the different formations it makes.

2. If your sand gets a little wet when you pull it out, squeeze it off in your hand and then let it dry on a paper towel or tray. It will return to its water-repellent state once dry.

PRO TIP: If you want to add a little extra fun to your sand, you can use an alcohol-based ink and an airbrush to color the sand before you go through the waterproofing steps.

FUN FACT: Magic sand is able to clump up in water because of the hydrophobic nature of the waterproofing spray. Once sprayed, the sand becomes hydrophobic as well, causing the sand to cluster, which minimizes the amount of space water has to penetrate the sand.

BOOM-erang! Using paint sticks and a little glue, you'll have a g'day with this craft.

PAINT STICK BOOMERANG

23

SAFETY KEY:
+ Sharp edges

SKILL LEVEL:
EASY
INTERMEDIATE
ADVANCED

APPROXIMATE TIME:
20 minutes

WHAT YOU'LL NEED:
+ Paint sticks
+ Wood glue
+ Sandpaper
+ Block of wood
+ Clamp or clothespin
+ Square measuring tool (or tool with 90-degree angle

G'DAY, MATE! Everyone likes a good ol' boomerang. Originating in Australia, it's a stick with a curve in it with some shavings and carving along the sides that, if thrown correctly, will wrap around and return to you.

LET'S BEGIN

BUILD YOUR BOOMERANG

1. Purchase the flattest paint sticks you can find. Test out if your stick is curved by eyeballing it or laying it against a flat surface. If it is curved, make sure that when you attach it the curvature is facing upward, forming a shallow bowl shape.

2. Find the exact center by measuring and marking both sticks and lining them up.

3. Add wood glue and use a clamp or clothespin to hold them together perpendicularly at the center mark. Use your square tool to check that each angle is 90 degrees and adjust as needed. Let the glue dry before removing your clamp or clothespin.

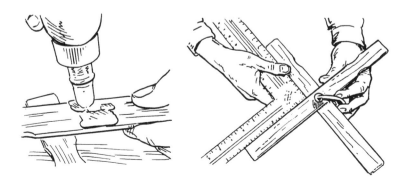

4. Using sandpaper, shape each arm like a wing (draw the wing lines with marker before sanding). A boomerang has a leading edge and a trailing edge. You throw it with the leading edge, and it comes back with the trailing edge.

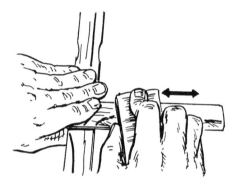

5. Sand down about half of the stick's trailing edge with a block of wood so it tapers down.

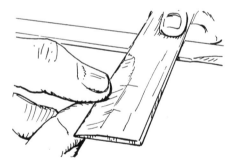

6. Round off the leading edge.

START BOOMERANGING!

1. Throw it straight up and down (vertically) and it should fly out and return to you. Each will fly a little differently, so experiment until you get it to boomerang just right. (Note: The wind will also affect the way it flies through the air.)

FUN FACT: From Egyptian kings to Australian Aboriginals, boomerangs have long been used as hunting weapons, the origins of which can be traced back to the Stone Age. They have been made out of bones, wood, and even mammoth tusks.

Have you ever dreamed of being a spy like James Bond or Jason Bourne? An old-fashioned staple of the spy is being able to encrypt and decode secret messages. Start your espionage now with this quick and easy recipe!

DIY INVISIBLE INK

24

SAFETY KEY:
+ Dangerous liquids

SKILL LEVEL:
[EASY]
INTERMEDIATE
ADVANCED

APPROXIMATE TIME:
15 minutes

WHAT YOU'LL NEED:

+ Paper
+ Lemons
+ Bamboo sticks/stylus pen
+ Electric stovetop
+ Paper towel
+ Spray bottle

+ Lighter
+ Baking soda
+ Grape juice
+ Cup or plate
+ Bleach
+ Ultraviolet flashlight

LET'S BEGIN

We are going to make three types of ink.

LEMON JUICE INVISIBLE INK

1. Squeeze a lemon (or buy pre-squeezed lemon juice from the store) into a glass or onto a plate.

2. Dip a bamboo stick in the juice.

3. Bring the stick over to a sheet of paper and slowly write a message. Re-dip in the lemon juice as often as needed.

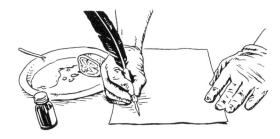

4. To reveal your message, carefully move a lit lighter back and forth under the paper, or move the paper over the heat of an electric stove top, making sure not to burn the paper. When the paper is properly heated, the message will appear.

ALSO TRY: Use an old-fashioned fountain pen to see how this affects the appearance of your mysterious messages.

BAKING SODA AND WATER INK

1. Mix equal parts baking soda and water.

2. Using the bamboo stick again, dip it in the mixture, and write your message.

> **PRO TIP:** Use a paper towel to blot excess liquid.

3. Put some grape juice in a small spray bottle and spray it on the paper. It tastes good as a juice; now let's see if it's a good secret-message revealer.

4. Your message should be revealed. How does this compare to the lemon juice?

BLEACH INK

1. Pour a small amount of bleach into a cup or onto a plate.

2. Dip your bamboo stick in the bleach and write a message.

In theory, the bleach is supposed to glow and be reflective under an ultraviolet flashlight. However, the paper around it actually glows and the bleach stays nice and dark.

3. Wave the ultraviolet flashlight over the message and it will appear.

The coolest part about the ultraviolet flashlight is that the message only appears there when it has the light on it, unlike the lemon and grape juice messages, which reveal the message permanently once you've gone through all the steps. With the ultraviolet flashlight method you can also see what's on the paper as you're writing it!

WHAT'S YOUR FAVORITE METHOD? Now that you've tried all three methods, think about which one would be your go-to method if you were actually a spy. You're one step closer!

FUN FACT: Invisible ink was first mentioned by Aeneas Tacticus in the fourth century BC in his book *How to Survive Under Siege.*

Throwback! If you remember the early '90s, then you'll remember slap bracelets—a super-popular accessory from back in the day. Now you can make them for yourself and rock some vintage style. They're also just fun to play with!

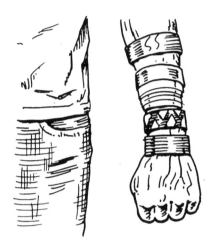

SNAP BANDS

25

SAFETY KEY:
+ Sharp objects—scissors, blade, screwdriver

SKILL LEVEL:
EASY
INTERMEDIATE
ADVANCED

APPROXIMATE TIME:
20 minutes

WHAT YOU'LL NEED:
+ Duct tape
+ Metal measuring tape
+ Scissors
+ Screwdriver (or any bar-like device with a smooth, round surface)
+ Bench vise

LET'S BEGIN

1. Extract the measuring tape, making sure not to let it spring back into the container. Also make sure you use a measuring tape that you are okay with ruining for this project.

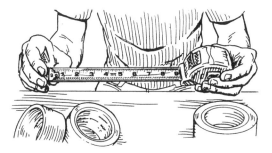

2. Cut the measuring tape into seven-inch lengths.

PRO TIP: Take out the whole measuring tape at once or tape the end onto its container, since once you cut off one piece, the remainder can be hard to manage.

3. Round off the edges and file down any sharp spots. The edges can accidentally cut you.

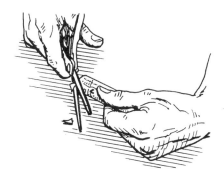

4. You need to take the curve out of the tape. Place a screwdriver in a bench vise and lock it down.

5. Bend the measuring-tape piece around the screwdriver, forcing it to bend the opposite way to take some of the curve off.

6. For safety, wrap colorful duct tape around the metal tape.

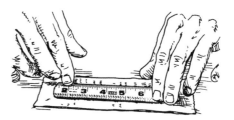

7. Use two pieces to cover it and trim off any excess. It should snap onto your wrist exactly how you want it to.

8. Decorate as you like. Use any color or pattern to personalize your snap accessories!

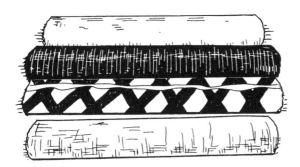

RANDOM THOUGHT: If we can make a slap bracelet, do you think we could also make a slap belt? A slap headband? Or any other accessory made out of this measuring tape, for that matter? Get creative!

FUN FACT: Snap bands were invented by a high school teacher in 1983. Ironically enough, they became so popular with students that some schools eventually decided to ban them, as they were causing distractions in class.

This spear-slinging device can be made in a short amount of time and with items you might already have around the house.

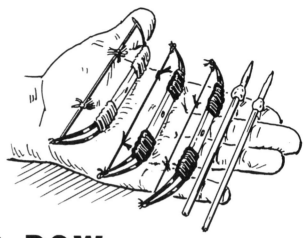

MICRO LONG BOW

SAFETY KEY:
  + Hot glue gun + Sharp knife

SKILL LEVEL:
  EASY
  INTERMEDIATE
  ADVANCED

APPROXIMATE TIME:
  10–20 minutes

WHAT YOU'LL NEED:
  + Bamboo skewers
  + Metal snap hair clips (barrettes)
  + Mini Popsicle sticks
  + Embroidery thread
  + Hot glue gun
  + Scissors or sharp knife

26

LET'S BEGIN

PREP

1. Make a hole the same circumference as the bamboo skewer in the center of a Popsickle stick: Use the point of a sharp knife and spin in circles, or a 9/16-inch drill bit to drill a hole.

2. Get rid of the center pieces in two hair clips by using your finger to press the back of the hair clip. When it pops up, grab that center piece and bend it over so that it snaps clean off. Press the center of the hair clip to pop it back into place. Do this to both hair clips.

CREATE THE JOINTS

1. Using a hot glue gun, glue the wide ends of the hair clips to the ends of the Popsicle stick.

PRO TIP: Put the Popsicle stick on a flat surface with the points of the clips facing up to allow it to dry straight. Wait about twenty to thirty seconds for the glue to cool and set completely.

2. You will reinforce the new joints (made from the hair clips) using the embroidery thread. Do each of the reinforcing steps on both sides.

3. Put a small dab of hot glue on the back side of the Popsicle stick (the opposite side from where you have glued the hair clips) and use that as a starting point to wrap your embroidery thread. Wrap the thread about five or six times around the base of the hair clip.

4. Once you have done that, start wrapping from just below where the hair clip is glued back up to the top of the hair clip until you cannot see where the Popsicle stick and hair clip overlap.

5. Use one more dab of hot glue to secure the end of the thread on the back.

> **PRO TIP:** If you lick your finger and press it down on the dab of glue, it will help prevent you from burning your fingers while also securing the thread even more.

MAKE THE BOWSTRING

1. Using the same embroidery thread, start with a piece of string that is a little longer than a bamboo skewer. Loop the thread through the centers of both hair clips, then bring it back down and loop the thread through the small hole at the top of the hair clip.

2. Make an overhand knot, tying it off at the small hole at the top of the clip. Put a small dab of hot glue there to keep it from unraveling. Tie it a second time before the glue cools to reinforce it.

3. Before you tie off the other hair clip, twist the thread to make it more durable.

4. With the string tightened, loop it through the small hole at the top of the other hair clip and tie it in a single knot. While tying it, you will want to press the bow against a flat surface so that you can tighten the bowstring before the knot is fully secured. Secure it with glue and a second knot.

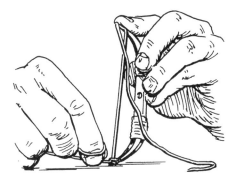

5. Snip off the extra thread at the ends to reuse as silencing tassels for the bowstring.

SILENCING TASSELS (OPTIONAL)

1. Tie the extra thread to the bowstring about one inch from the bottom using a double knot. Do this at both ends of the bowstring.

2. Snip them down so that they are about half an inch long and ruffle them up.

MAKE THE ARROWS

1. Find the straightest bamboo skewers. Measure the bamboo skewer against the bow. You want your arrow to be about as long as the bow. Break off the part of the skewer that is longer than the bow.

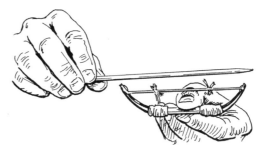

2. To make the arrows fit better into the bowstring, make a small notch at the bottom of the skewer using a pair of scissors or a sharp knife.

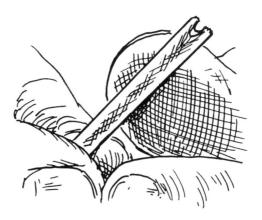

3. Add a dab of glue about an inch down from the tip to help the arrows fly through the air. Rotate the glue gun around that area to make sure the glue goes around the skewer.

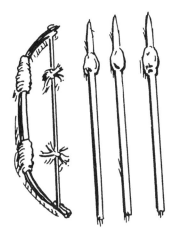

SHOOT!

Now that you're all done, you can try shooting the arrows at objects at short range. We tried shooting it at a piece of toast that we stood up and it worked great. Enjoy!

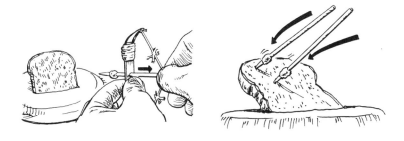

FUN FACT: Originally called the "war bow," this weapon dates all the way back to the prehistoric era. It gained popularity in England later on and became a popular fighting tool in medieval times.

INTERMEDIATE
PROJECTS

Take the technology behind a crossbow and miniaturize it to create a compact, powerful skewer shooter that can fit in your shirt pocket. To kick it up a notch, add a custom-fitted leather bandolier to your backyard barrage!

ARM-MOUNTED SKEWER SHOOTER

27

SAFETY KEY:
+ Never use contact cement where you do not have good ventilation. It can cause respiratory tract, eye, and skin irritation. Never shoot at people, animals, or property.

SKILL LEVEL:
EASY
INTERMEDIATE
ADVANCED

APPROXIMATE TIME:
1 hour

WHAT YOU'LL NEED:
+ Pack of cheap disposable pens
+ Thick rubber bands
+ Strap of leather
+ Snap-making kit
+ Bamboo skewers

+ 12 (1×1) LEGO hooks
+ 3 LEGO plates
+ Contact cement
+ Sandpaper
+ Pliers

LET'S BEGIN

SETUP

1. Remove the cap from a disposable pen and use a pair of pliers to pull off the front and back pieces. Use glue to reattach the back stopper to where it was in the back of the pen. Add a tiny drop of superglue around the base.

2. Hold a thick rubber band on the top end of the pen and pull the band down toward the end piece. Mark on the pen where you want to attach the rubber band.

ATTACHING THE RUBBER BAND

1. Place the rubber band on the pen where your mark is, so the bottom loop is by the closed cap. To make sure the rubber band lines up with the pen as you attach it, cross the two sides of the band over and wrap them around the back of the pen, which will help the rubber band lie flat against the pen as you tape it.

2. Wrap a piece of strong tape very tightly down against both the rubber band and the body of the pen.

3. Add a few more layers of tape to the top of the pen to make a finger grip.

PRO TIP: Now's a good time to color any part of the pen and rubber band; you want to give it a sleek, professional look.

MAKING AND FIRING BAMBOO SKEWERS

1. The skewers fly best when weighted, so add some tape to the tip.

2. To fire, simply load the skewer into the barrel so the back presses against the rubber band, pull back, and let her rip! You should be able to shoot up to seventy feet.

MAKING THE LEATHER BANDOLIER

Take your firing capabilities to the next level with a custom armband bandolier to hold your arrow supply. Now you can shoot multiple arrows in a row without having to pick them up first.

1. Wrap a piece of leather around your arm and mark on the leather where it fits snugly on your arm. Add about one and a half inches to that and make your cut.

2. Drill a hole on each end of the strap where your snap will go.

PRO TIP: Place a wooden block under the leather to protect your table when drilling.

3. Using the tools provided in the snap kit, you can easily secure the metal pieces into the leather. With both sides attached, the armband should fit perfectly.

1. Use a dab of glue to attach twelve LEGO hooks to the ends of three LEGO plates. Slightly offset the hooks so the skewers can't slide freely through.

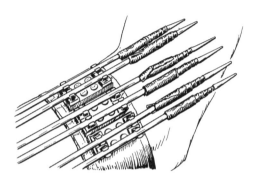

2. Mark on the bandolier where you want the LEGO plates to go. Then rough up the spot with sandpaper so the glue has something to hold on to.

3. Paint a thin layer of contact cement onto the leather and the bottoms of the LEGO plates. When the glue no longer feels sticky to the touch, it's time to press the plates to the leather, forming an incredibly strong bond. Snap in your arrows, and you're ready for action!

All in all, this little arrow launcher costs about $3 to make and for just a little bit more you can make yourself a cool armband that holds all of your ammunition. These wooden darts can pop balloons, travel over seventy feet, and even pierce targets designed for metal tips.

FUN FACT: Bamboo releases 30 percent more oxygen into the atmosphere and absorbs more carbon dioxide compared to other plants.

Turn household items into a
micro crossbow that accommodates
multiple rounds of ammunition,
launches exploding tipped crossbow
bolts, and slings wooden matches
over thirty feet away!

ASSASSIN'S MICRO CROSSBOW

28

NOTE: This project was in
collaboration with the Sonic
Dad team, with credit to
Ritchie Kinmont for designing
the template. Check them out
at www.sonicdad.com

SAFETY KEY:
+ Use outdoors + Adult supervision recommended

SKILL LEVEL:
EASY
INTERMEDIATE
ADVANCED

APPROXIMATE TIME:
1 hour

WHAT YOU'LL NEED:
+ Popsicle sticks
+ Scissors or gardening
  shears
+ 2 metal snap hair clips
  (barrettes)
+ Embroidery floss
+ Hot glue gun
+ Sharpie

OPTIONAL:
+ Matchsticks
+ Pop-Its
+ Electrical tape

LET'S BEGIN

MICRO BOW

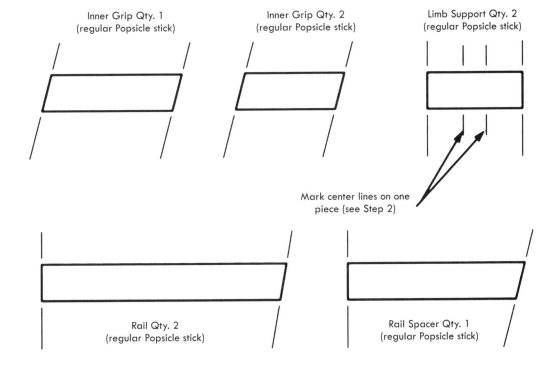

1. Use the template below to mark your Popsicle sticks and trim them to size.

Inner Grip Qty. 1
(regular Popsicle stick)

Inner Grip Qty. 2
(regular Popsicle stick)

Limb Support Qty. 2
(regular Popsicle stick)

Mark center lines on one
piece (see Step 2)

Rail Qty. 2
(regular Popsicle stick)

Rail Spacer Qty. 1
(regular Popsicle stick)

2. Color each of the sticks black using a Sharpie.

PRO TIP: Using a large black Sharpie is the quickest and easiest way to get it done.

3. Find two metal hair clips, and break the inside band cleanly out of the center. Secure the clips to the smallest wooden supports (cut Popsicle sticks) with a little hot glue. When it cools, the limbs should arch away from each other in a bow shape.

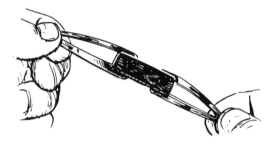

4. Add some hot glue to the inside and press the second support firmly into place, then when it cools, clean up any excess glue with a utility knife.

MAKE THE PISTOL GRIP

1. Use hot glue to join a crossbow rail to the inner rail spacer so they're matched flush with the tip.

PRO TIP: Drop the inside rail down a touch to form a little groove about half a matchstick deep, then remove any excess glue to ensure that the flight channel stays clean.

2. Attach the handle support right behind the spacer, then glue the last rail in place on top.

HELPFUL HINT: Leave a matchstick in the flight channel to make sure the glue cools with enough spacing.

3. Finally, glue the pistol grip panels so they overhang slightly at the back.

PREP FOR ATTACHMENT

1. You'll need a shallow groove for catching the bowstring, so whittle out a small catch in the upper rails, just above the front of the pistol grip.

PRO TIP: Round the top edges of the flight channel with sandpaper to prevent the bowstring from fraying and breaking later on. Don't worry about messing up the paint job—it's super easy to touch up with a marker and make good as new.

2. Connect your pistol grip to the bow by placing a bead of hot glue on the inside of the bow and pressing in the tip of the gun rail. Make sure the inner rail is flush with the top of the bow.

3. Add glue to each side of the rails for added support and durability, then give it a couple of minutes to cool and harden.

STRINGING THE CROSSBOW

1. The crossbow strings are made with embroidery floss. Thread the floss through the circular holes in the tips of the hair clips and tie each one with a double knot. Add a bit of glue to keep the knots from unraveling.

PRO TIP: Twisting the bowstring before tying it off will help keep it tight. This keeps tension on the string and increases the draw weight of the bow.

HELPFUL HINT: You can find embroidery floss in the craft section of supercenters like Walmart.

2. Before moving on, it's a good idea to reinforce the bow by adding string where the hair clips meet the grip. Glue your string to the back side of the bow and wrap the support around ten times on each side.

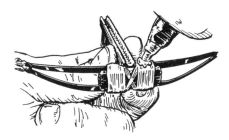

3. Finish by adding some glue to this knot as well, and trim the excess thread.

ALSO TRY: For bonus points you can wrap string around the handle to create a custom grip.

FINISHED! TIME TO FIRE YOUR CROSSBOW

1. Cock the bowstring by drawing it back into the notch you made earlier.

2. Insert a wooden match into the groove, and use your thumbnail to press up gently on the string.

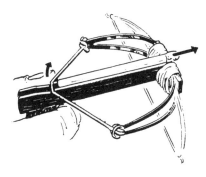

3. The string will snap into the back of the match, sending it flying off at incredible speeds and up to thirty feet away!

PERFECT IT! You can increase effectiveness and accuracy by using scrap wood to create a retention spring on the back.

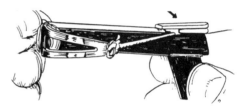

OTHER THINGS TO TRY

SIDE-MOUNT QUIVERS: Cut two small pieces of a drinking straw, cap one end on each with hot glue, and attach them to either side of the barrel at a bit of an angle.

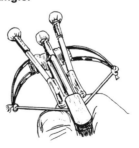

EXPLOSIVE BOLT HEADS: Flying matches are cool, but what if they exploded on impact? Secure an individual Pop-It to the tip of a match using electrical tape. Now your bolts will make a bang when they hit a hard target.

TARGETS: Make your own targets to practice your accuracy. You could make ones using cardboard and glued-on images or anything that your creative mind can come up with.

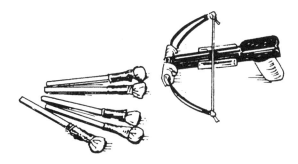

BE INNOVATIVE! Whether you stick to simple crossbow game play or take it to the next level with explosives, there's always more to explore with this versatile weapon in your arsenal.

> **FUN FACT:** Crossbows may seem like a noble weapon now, but back in medieval days, the weapon was seen as so dishonorable that crossbowmen were paid double the amount of a regular soldier to use them.

Simple to make but powerful enough to break glass and blast darts into concrete! In this project we're making a custom laser-guided blowgun that not only looks cool but adds another dimension to your backyard endeavors.

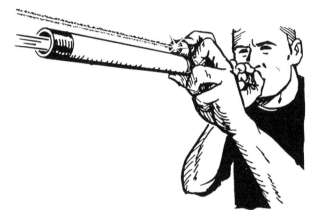

LASER-ASSISTED BLOWGUN

29

SAFETY KEY:
+ Adult supervision + Open flame + Sharp projectiles
+ Safety shades + Danger: Concerned neighbor warning
+ Check legalities

SKILL LEVEL:
EASY
INTERMEDIATE
ADVANCED

APPROXIMATE TIME:
30 minutes

WHAT YOU'LL NEED:
+ 2-foot length of $1/2$-inch PVC pipe (Schedule 40)
+ Female reducing adapter (Schedule 40, $1/2$ inch × $3/4$ inch)
+ Zip ties
+ Laser pointer
+ 2 × 2-inch sticky notes
+ Paper party hat
+ Scotch tape
+ Hot glue gun

+ Blue tack adhesive putty
+ Sharpie

OPTIONAL:
+ #16 and #18 wire nails
+ Duct tape
+ Electrical tape
+ $1/2$-inch-thick foam pipe insulation

LET'S BEGIN

MAKE THE DARTS

1. Wrap the sticky note around the tip of the party hat to form a pointed cone, and secure with Scotch tape.

HELPFUL HINT: Use two small pieces of tape about one inch each and be careful you don't tape your paper dart to the hat.

2. Push the tip of the dart into the ½-inch PVC tube and give it a little twist. The pressure will leave an indentation right where the dart needs to be cut to fit inside the tube. Use scissors to carefully trim your dart along the line.

3. In order for your darts to fly straight, you'll need to add some weight to the tips. Try squirting a little hot glue inside the cones until they're about a third of the way full, or use a screwdriver to shove a small piece of adhesive putty down into the tip.

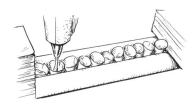

4. Now you've got a lineup of mini darts that should fit perfectly inside the barrel and fly impressively straight.

PRO TIP: You can take it one step further and transform your darts into high-speed nail darts. Simply push a nail through the cones until the heads catch on the inside. Then add a dab of hot glue on the inside to make it more durable.

HELPFUL HINT: Feel free to experiment with different kinds of heads, weights, and lengths. Try using #16 and #18 wire nails because they are relatively lightweight and have a nice flat head on top. Smaller darts will be easier to shoot and fly faster, while heavier darts will go slower but will penetrate deeper and do more damage.

ASSEMBLE YOUR BLOWGUN

1. Connect the reducing adapter to your half-inch PVC pipe, and tap gently. Just like that, you have a super-simple blowgun that can be used right away!

CUSTOMIZE YOUR BLOWGUN WITH COLOR AND LASERS

1. Get a roll of duct tape with a pattern you like, and lay out a strip on the table with the sticky side facing up.

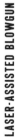

2. Position your pipe lengthwise in the center of the strip and gently rock it back from side to side so the tape wraps itself around the tube. Use a second piece of tape on the other side to cover any gaps.

PRO TIP: Use electrical tape around the tip of the barrel and the mouthpiece to cover up any rough edges, and give your blowgun a clean, finished look.

STEP UP YOUR GAME: Take it to the next level by adding a laser sight! Hot-glue a dollar-store laser about eight and a half inches from the end of the barrel. Then hold it in place with two zip ties.

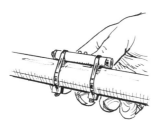

ONE STEP FURTHER

You've got a custom laser blowgun, an arsenal of darts, and everything you need to begin target practice. But why not take it to new heights by building an adjustable scope and a quiver?

1. Cut off a three-inch length of PVC, and buy a more powerful laser.

2. Cut a hole in the top of the PVC so you have access to the button. Place the laser inside and secure on either end by driving four small screws into each end. Then secure it to the blowgun with zip ties.

PRO TIP: You can calibrate your scope by securing the blowgun in place with something like a bench vise, taking a shot, and adjusting the screws on the casing until the laser points to the exact spot where the dart hit.

3. To form a quiver, cut two ¾-inch-wide pieces off the end of your foam pipe insulation.

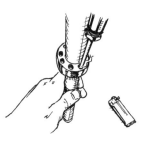

4. Wrap the outsides of the disc with electrical tape, and color the rest of the foam with black marker, just for looks.

5. Slide the foam discs onto the blowgun barrel. Use a lighter to heat the tip of a Phillips-head screwdriver for about twenty seconds. Then make eight evenly spaced holes in the foam disc on the top and slide in the darts.

This quiver allows your darts to slide into place, and helps prevent the tips from poking out.

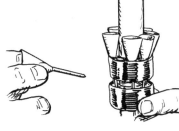

LASER-ASSISTED BLOWGUN

FINISHED! LET'S PLAY

These simple darts will penetrate dartboards, wooden posts, tree trunks, and even cinder blocks! With some practice you should be able to hit a target from one hundred feet away.

For a safer kind of ammo, try mini marshmallows. They work perfectly!

☠ **WARNING:**
Even though they're made of household materials, these darts pack a punch. Never shoot at people, animals, or property that isn't yours.

FROM TRASH TO ARSENAL TREASURE: This project came to be by simply taking a plastic pipe out of the garbage and imagining what its use for fun could be. Eventually it became an all-encompassing dart gun that fires at high velocity! You never know when your junk will become your next favorite desktop weapon. There's so much rubble just waiting to be developed into the next cool device.

FUN FACT: When you think of laser tag, images of preteens running around may come to mind. However, the origin of this game is way more serious: it was created in the 1970s for US Army soldiers to train for combat in a nonlethal way.

This Skyblaster Slingshot will lob water balloons over 150 feet, shoot up to three balloons at a time, and wreak havoc in your next water fight!

SKYBLASTER SLINGSHOT

30

SAFETY KEY:
+ Use outdoors

WARNING:
☠ The balloons hit hard enough to leave welts, and using a person as a target is not advisable.

SKILL LEVEL:
EASY
| INTERMEDIATE |
ADVANCED

APPROXIMATE TIME:
2 hours

WHAT YOU'LL NEED:
+ $1/2$-inch PVC tubing
  Two $1 3/4$-inch pieces
  Two 2-inch pieces
  Four 5-inch pieces
  Two 10-inch pieces
+ Sprinkler fittings
  Two $1/2$-inch PVC slip clip
  Two $1/2$-inch PVC tees
  Two $1 1/2$-inch 90-degree PVC elbows
Four $1/2$-inch 45-degree PVC elbows
+ PVC quick-dry cement
+ 50-inch resistance exercise band (with handles)
+ Disposable pen
+ Spray paint or duct tape
+ Zip ties
+ Water balloons (for ammo)

PRO TIP: These adapters are all standard, and should be available at a home-improvement store or sprinkler-supply outlet near you.

LET'S BEGIN

BUILDING YOUR FRAME

1. Once your PVC tubing is cut to size, dry-fit the pieces together before gluing, using the images in this project as a reference for the shape of your frame.

2. Use PVC quick-dry cement to glue the parts together **except** for the two inside five-inch tubes.

3. Customize your bow frame with spray paint or duct tape.

4. Use scissors to cut off the resistance-band handles and pull off the foam exteriors. Twist the foam grip onto each of the two five-inch tubes.

5. Glue the two five-inch tubes onto the frame to complete it.

1. Connect the ends of the fifty-inch workout band to make one rubber circle. To do this, take apart a plastic disposable pen. Cut two to three inches off the barrel of the pen and then use a scissor blade to score 45-degree markings up and down the sides. Push both sides of the pen into the ends of the tubing until a circle is formed.

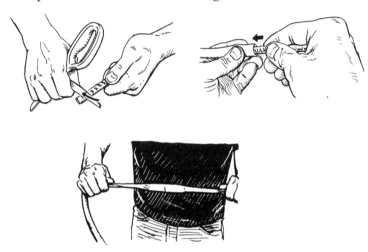

2. To add your launching pouch, cut duct tape into four strips that are eighteen inches long and four strips that are fourteen inches long. Connect your four longer strips together by overlapping them horizontally, giving you a nice wide piece of duct tape fabric.

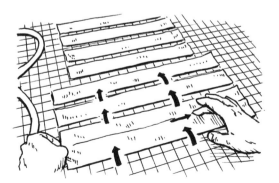

3. From the edge, measure the following marks: five inches, two inches, four inches, two inches, five inches. In the center of each of the two-inch marks, place the fourteen-inch pieces of duct tape facedown and press them into place, leaving you with a four-inch section in the center and five inches on either side. Wrap the excess ends of the fourteen-inch pieces around to the back side and press smooth.

4. Bring back the big rubber band and stretch it sideways, so the band lies exactly on top of the black strips, making sure that the pen-tube connection is in the center. Attach the band by folding the duct tape over the band and securing it in the center of the circle.

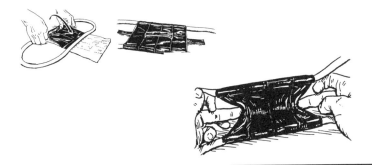

5. Grab the last two strips of duct tape. Fold one of the strips lengthwise into thirds. Wrap the other piece of tape around it to strengthen it.

6. Cut two slits in the center of the pouch two inches from either end. Weave the tape strap through both holes. Trim the ends of the strap so when folded in toward the middle, they meet the edge of the handle in the center. Secure with tape.

ATTACHING THE LAUNCHER TO THE BOW

1. Line the ends of the tubing up with the grooves in the tips of the bow. Stretch the rubber down as hard as you can, rocking it back and forth so it slips inside the groove. Use a four-inch zip tie to pinch the tubing tightly together. Repeat for the other side.

 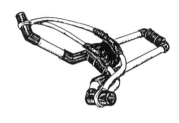

LOADING AND FIRING

1. Load your slingshot by placing a water balloon in the pouch while it's facing upward.

2. Pull back to lock onto the balloon, then fire when ready. With only a bit of effort, you will be able to send balloons high into the sky and up to 150 feet away. To get the best leverage, try shooting it like a bow. As you practice, your shots can get incredibly powerful and impressively accurate as well.

ALSO TRY: If you don't feel like getting wet, forget the water balloons and try shooting Sky Ballz (chapter 13) instead!

READY FOR BLASTING! This extreme slingshot only costs about $10 in materials, and is also durable, customizable, and powerful. It'll give you an edge over the competition while you're defending and dominating in your next water fight!

FUN FACT: Slings have been used for millions of years, but the slingshot is relatively new on the scene. Developed in Russia, the slingshot became popular with children, as they could be easily made from wood and rubber inner tubes.

Whether you just want to play with flames or need to get a campfire going in a survival situation, this cheap and easy tool will allow you to harness the full force of the sun into a powerful pinpoint.

MINI SOLAR SCORCHER

31

SAFETY KEY:
+ Fire

SKILL LEVEL:
EASY
INTERMEDIATE
ADVANCED

APPROXIMATE TIME:
30 minutes

WHAT YOU'LL NEED:
+ 10 paint sticks
+ Fresnel lens (you can get one at the dollar store)
+ Piece of cardboard
+ Everbilt hardware:
  #8-32 × $\frac{1}{2}$-inch machine screws combo
  #8-32 × $\frac{3}{4}$-inch machine screws combo
  #8-32 × $1\frac{1}{4}$-inch round head combo

#8 flat washers
#8-32 wing nuts
#8-32 × $1\frac{5}{8}$-inch eye bolts
+ Wood glue
+ Drill
+ Aluminum foil
+ Saw
+ Pliers

LET'S BEGIN

BUILDING YOUR FRAME

1. Use eight paint sticks to build two wooden frames that will sandwich the Fresnel lens. Cut each of the sticks at a 45-degree angle on both ends. The exact length depends on the size of your lens, but you'll probably end up making a rectangle. After you cut the angles off the paint sticks, you will have eight little triangle scraps. Save those for later.

PRO TIP: If you don't have a tool to measure a 45-degree angle, fold a square piece of paper diagonally.

2. Use wood glue to construct two wood rectangles. While you're at it, glue the saved wood triangles onto the frame corners.

3. Glue the Fresnel lens between the two rectangles, pressing tightly.

4. Drill three equidistant holes through each side of the frame. Thread #8-32 × ½-inch machine screws through the top and bottom holes of each side.

5. Drive 1¼-inch machine screws through the middle holes on the short side. These will be your sun finders, so they need to be as straight as possible.

6. Attach two eyebolt screws to the middle holes on the long sides of the frame. Secure each with a bolt, making sure these are on the rough side of the lens.

7. Drill a hole through the wide ends of four paint sticks. Add a screw and a washer to the end of the eyebolt, then add a paint stick. Follow up by adding two washers, another paint stick, and then end with a washer and wing nut. Do this to both sides.

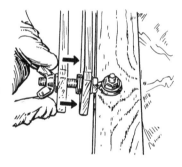

8. Now you have two pieces of wood hanging from each side, and it's time to connect them. Measure the distance between the opposing legs. Cut four sticks to size and glue them across, creating a reinforced stand and a shelf.

9. Cut a piece of cardboard to fit over the lens and a piece of tinfoil to cover the cardboard. Glue into place.

USING YOUR SOLAR SCORCHER

1. Take it outside and find a sunny spot. Choose what you want to burn—try starting with a spare paint stick. Hold it underneath the lens and slowly raise it until you find the center point. You will know you've hit focus when your stick instantly catches fire!

ALSO TRY: Using the tinfoil–covered cardboard, you can add colored powder and slide it slowly out from the bottom until it ignites. Try playing around with smoke powder, a book of matches, two-by-fours, or anything small. But whatever you do, don't hold small things in your hands.

From cracking fireworks to fireballs on burning matchsticks, you'll be impressed with the potential of this small but mighty scorcher.

> **FUN FACT:** Photons are the particles that bring light from the sun to the earth. When narrowly focused, the energy from the photons can light things on fire once it reaches 450 degrees Fahrenheit.

Turn a ninety-eight-cent mousetrap into a fun little handgun that can shoot projectiles with force, precision, and a satisfying little kickback!

MOUSETRAP GUN

32

SAFETY KEY:
+ Safety glasses

SKILL LEVEL:
EASY
INTERMEDIATE
ADVANCED

APPROXIMATE TIME:
30 minutes

WHAT YOU'LL NEED:
+ 2 pack Tomcat mousetraps
+ 2 × 2-inch wood block
+ 2 Phillips screws
+ Needle-nose pliers
+ Drill with $\frac{1}{8}$-inch bit

+ Ammunition (airsoft BBs)
+ Sharpie

LET'S BEGIN

BUILDING YOUR HANDGUN

1. Cut your piece of two-by-two to the height of four fingers. This will be your handle.

2. Remove the plastic yellow bait pad from a mousetrap. Use your pliers to pull out the industrial staple from the base.

3. Place the trap over the handle and use a Sharpie to mark two spots to drill holes to fasten the trap to the handle. Once marked, safely place the trap on your workbench and use an ⅛-inch bit to drill holes all the way through the dots. Mark where your handle will be placed. The back of the trap should overhang the handle by a quarter of an inch. Then mark a third spot to drill a hole that is slightly in front of the handle. This will be for the locking pin trigger.

4. Use two screws to fasten the trap to the top of your handle, screwing through the holes on the heel bone and elbow.

5. Shorten the locking pin by holding it over the spring, looking down from the top, and marking where it lines up just past the remaining hole. Use the snips on your pliers to cut to size. Bend the tip back 45 degrees.

6. Remove the locking pin from the second trap and thread it through the remaining hole. Trim the pin so it's flush with the spring and then bend it over, forming a hook.

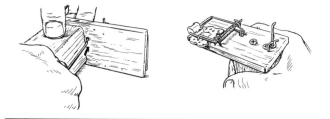

7. Turn the yellow bait pad into a launchpad by clipping it onto the trap hammer with its hooks facing up. Slide it all the way to the right side, then simply lift the hammer up and tuck the pad inward so it lies down flat on the platform.

PERSONALIZE IT! Paint your trap your favorite color or use duct tape to create a custom grip on the handle.

HOW TO FIRE

1. Setting the firing mechanism works about the same as setting the trap, only this time you're pushing the trigger up from the bottom so the hook catches the locking pin in place. It's a three-step process to pull the hammer back, set the pin, and secure it with the trigger hook.

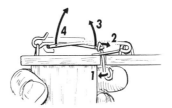

2. Place your ammunition on the launching pad. Airsoft BBs fit perfectly in the circular holes. Pennies and small rocks work great as well.

3. Squeeze the trigger gently, causing the pin to slip and the launching pad to snap forward, springing your ammo straight ahead.

PRO TIP: If you're feeling lazy just pull back the launchpad with your thumb and release when you're ready. This opens up the option for rapid firing.

☠ WARNING:
When firing make sure you keep your face back from the gun. The locking pin will snap back and could injure you.

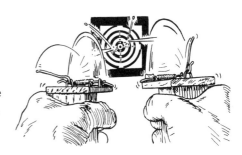

There you have it! All you need to convert two mousetraps into a mini handheld catapult are some simple tools, two screws, and a small scrap of wood. It won't keep your house rodent-free, but it is a cheap, easy way to have some spring-loaded fun.

FUN FACT: Over 4,400 patents have been awarded to mousetraps. This is more than any other device, and according to the Smithsonian, the race to build a better mousetrap symbolizes the American drive to innovate.

Indiana Jones has nothing on this cheap, sturdy, and seriously striking bullwhip!

PARACORD BULLWHIP

SAFETY KEY:
+ Safety glasses

SKILL LEVEL:
EASY
INTERMEDIATE
ADVANCED

33

APPROXIMATE TIME:
90 minutes

WHAT YOU'LL NEED:
+ ³/₄-inch PVC
+ 1-inch PVC
+ Paracord
+ Daisy 2400-count BBs
+ Athletic tape
+ Electrical tape
+ Drill
+ Lighter
+ Bench vise

LET'S BEGIN

MAKE THE HANDLE

1. Cut your ¾-inch PVC to the length that you want for your handle (nine inches is generally a good length). Also cut a one-inch-long piece of the one-inch PVC.

2. Push the smaller one-inch piece all the way onto the longer piece.

MEASURING

1. Run the paracord through the handle and measure eight feet in front of the handle and about two feet behind. It should be about ten feet in length. Use electrical tape to mark the bottom of the handle at the two-foot mark.

2. The next piece will run the length of the whip, and the last eighteen inches will be the "fall," which gives it that distinctive crack. Place another piece of electrical tape to mark the beginning of the fall. Cut another piece that measures the length between the two pieces of electrical tape.

3. For the remaining seven pieces, remove an additional eight inches of length for each cord, taking off another eight inches more with each piece. You should have ten pieces in diminishing lengths. If you're off by an inch or two for some of your pieces, don't worry; it won't make a big difference.

ADDING SOME WEIGHT

Traditional bullwhips are made of kangaroo leather, so you're going to need to add something extra to give your paracord whip more heft.

1. Cut three pieces of paracord that match three existing pieces (longest, shortest, and the middle piece is a good way to go). Pull out the white internal strands from the paracord completely. You can throw these away.

2. Seal off one side of each of the three cords by melting it closed with a flame.

3. Insert a ³⁄₁₆-inch drill bit into the open end of one of the strands. Melt the strand around the drill bit so it stays open. Repeat with the other two cords.

4. Fill each strand with BBs. This may take a while, but you'll finish with three strong yet flexible cords. Melt the ends to seal.

ASSEMBLING

1. Using electrical tape, secure all thirteen strands of paracord together.

2. Drill a ³⁄₁₆-inch hole three-quarters of an inch from the end of the handle. Thread the extra piece (the first piece that we cut) of paracord through until it pulls the entire bundle into the handle. Pull the strand tight until it pulls the rest of the pieces up through the handle. Take the remainder of the first piece and wrap it around the handle. Secure the cord on both ends of the handle with an ample amount of electrical tape.

3. Wrap a piece of electrical tape around the paracords about every foot and a half so they don't get twisted.

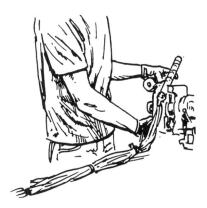

4. Secure the handle in a bench vise and carefully wrap the paracord bundle in athletic tape, keeping the strands tight and parallel. Start at the base of the handle until you reach the fall, but make sure you leave the fall exposed.

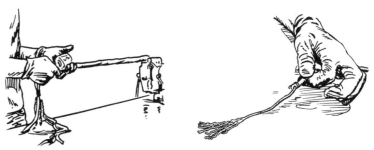

5. Now that you're done assembling, use colorful tape to decorate your handle.

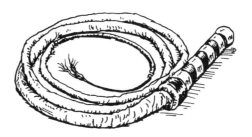

GET CRACKING

Time to take your whip outside and give it a whirl. The "Cattleman's Crack" involves hurling the whip over your shoulder. For the "Overhead Crack," spin the whip in a circle over your head before spiking it down in front of you. Just remember to wear long sleeves and safety glasses.

You'll become hooked by its sonic boom as you play around with more innovative ways to utilize your whip! Watch out, cowboys and movie stars, this earsplitting whip is a force to be reckoned with.

FUN FACT: The crack of a whip breaks the sound barrier as the tip zooms at up to 750 miles per hour. It is believed that a whip is the first man-made object to ever break the sound barrier.

At your next outdoor party don't be a wallflower! Grab someone's attention or charm the crowd with this colorful and customizable party whistle.

BOTTLE CAP PARTY WHISTLE

34

SAFETY KEY:
  + Sharp objects

WARNING:
  ☠ Cutting aluminum cans will give the metal very sharp edges. Sharp edges cut skin. You may want to use gloves to mitigate any risk.

SKILL LEVEL:
  EASY
  INTERMEDIATE
  ADVANCED

APPROXIMATE TIME:
  20 minutes

WHAT YOU'LL NEED:
  + Empty aluminum soda or beer cans
  + Bottle caps
  + Hot glue gun
  + Scissors

LET'S BEGIN

BUILDING YOUR WHISTLE

1. Use scissors to cut the top and bottom off an aluminum can, leaving you with a lightweight sheet of metal.

2. Cut a long narrow strip off the end of the metal sheet, the width of a single can tab. Then cut a smaller, stubbier piece as wide and as tall as two tabs laid side by side.

3. Join the two pieces together with the smaller rectangle centered on top, about a quarter inch down from the end of the long piece to form a lowercase "t."

4. Wrap the two sides of the small rectangle around the back of the long piece. Snip the top corners off and fold the tab over to lock everything together. This will be the mouthpiece of the whistle.

5. Now take the bottom end of the strip and push it back through the mouthpiece.

6. Cut a small notch in each bottle cap so when the two bottoms are lined up, they form a round container with a symmetrical, rectangular hole at the top.

7. Bring back the mouthpiece and fit the barrel inside one of the caps, expanding it to completely fill the inside of the cap. Line the inside of the cap with some hot glue, then place the rounded strip back in position.

8. Cut the extended strip where it lines up with the top of the bottle cap notch and then pull the scrap part out of the mouthpiece.

9. Add hot glue to the other cap and press the two pieces firmly together, making sure the notches line up.

10. Cut a can tab in half and hot-glue it to the back of the whistle. Now you have the option of adding a lanyard!

11. Put the whistle in your mouth and let 'er rip! The whistle works by directing air through the mouthpiece and splitting it over the sharp metal edge. If you try rolling your tongue when you blow, it will sound a lot like a referee's whistle and be ear-piercingly loud.

Show up to your next party loaded with one of these bad boys! It'll give you the power you need and the attention you deserve.

FUN FACT: The average whistle has a noise level of 104 to 116 decibels. To put that into perspective, a jackhammer has a decibel level of 100 and a thunderclap rings in at 120 decibels.

Got an itch to grill but don't have a barbecue handy? Here's an awesome little hack to satisfy your cravings one bratwurst at a time!

BITTY-Q

35

SAFETY KEY:
+ Fire + Sharp edges

SKILL LEVEL:
EASY

INTERMEDIATE

ADVANCED

APPROXIMATE TIME:
30 minutes

WHAT YOU'LL NEED:
+ Large drink can (about 1.5 pints)
+ Wire coat hanger
+ 60-grit sandpaper
+ Wire dikes
+ Pliers
+ Two 1-inch utility hinges
+ #4 round-head machine screws ($^3/_8$ inch)
+ #8 machine screws
+ Two 4-inch U-bolts
+ Strap loop
+ Sharpie

LET'S BEGIN

CUT THE CAN

1. Use scissors to cut the can cleanly in half lengthwise.

2. Make two small angled cuts about half an inch from the top of each sharp edge, which will allow you to form a small ledge. Try using a Popsicle stick as a guide so you can get a nice clean crease. When you've finished one side, go ahead and do the other side the exact same way.

SAFETY FIRST: Don't forget to trim down the pointy parts so they are not as sharp.

ASSEMBLING THE GRILL

1. Grab a wire hanger and a pair of wire dikes. Cut the hanger at the neck, then remove the protective coating using a piece of 60-grit sandpaper.

2. Using the image at right as a reference and a pair of pliers, measure and carefully bend the wire into the shape of a grill rack—giving you a removable grill that springs into position when you're ready to start cooking.

3. Remove the metal plates on two four-inch U-bolts. Place one plate on the outside of the can and mark the holes with a Sharpie. Use a pair of scissors to pierce the can in the four spots you marked, making sure the holes are big enough for the U-bolt.

4. Push the U-bolt through the holes and reattach the plate. Bend the two legs away from each other and the plate will lock down, holding the legs secure. Repeat with the other U-bolt and plate and toss the nuts back on the ends for decoration.

EXTRAS

1. You could stop here, but let's take it to the next level by adding a top. You'll be using a one-inch utility hinge to connect the pieces. Simply poke four holes on each side and use #4 round-head machine screws and nuts to fasten together, making sure the hinge barrels are facing inward.

2. Finish up by adding a miniature handle. Use two #8 machine screws to attach a strap loop to the edge of the lid and your Bitty-Q is completely finished and ready for grilling.

GRILL, BABY, GRILL

1. Simply toss in some charcoal, or whatever else you use for grilling. Snap the wire grate into position and ignite the fuel.

2. Add your food and close the lid to get everything sizzling. While you wait you can keep your tongs in the built-in loop provided by the piece of wire extending from under the lid.

3. In ten minutes or so your treat should be grilled to perfection and ready to satisfy those cravings.

GET CREATIVE: This is a great project to play around with! Try deploying two smaller cans as a base, or using a piece of wire lath instead of a clothes hanger.

If you find yourself without a grill and have an itch to cook up something tasty, you'll now know how to make an itty-bitty barbecue! It'll help you get the fix you need and put some sizzle on those links one glorious dog at a time.

> **FUN FACT:** Surf and turf today is usually steak and shrimp, but back in 29000 BC surf and turf might have been mammoth ribs and conch meat! In 2009, archaeologists unearthed a cooking pit with these two ingredients, suggesting that this famous dish may have been around longer than originally thought.

Put on your gloves and get ready to experiment with a pull-tab pyrotechnic. We will be making a fuse igniter from a book of matches so that you can start a fuse with a simple flick of the wrist.

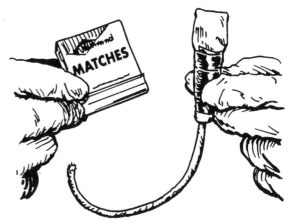

MATCHBOOK FUSE IGNITER

36

SAFETY KEY:
+ Fire

SKILL LEVEL:
EASY
INTERMEDIATE
ADVANCED

APPROXIMATE TIME:
15 minutes

WHAT YOU'LL NEED:
+ Book of paper matches
+ Electrical tape
+ Fuse

LET'S BEGIN

SETUP

1. Open up your book of matches to reveal the four layers of matches inside. When you grab onto the base and pull on the cardboard, the paper box should detach from the matches.

2. When you look at your matches, you will see a staple holding them together. You can use your fingernails, teeth, or perhaps pliers to remove that staple.

3. Next, lay the cardboard the matches were in down on a table so the outside packaging and striker strip are facing upward.

4. Next, take the outside edges and fold them toward the center until they meet in the middle.

5. To hold them together, take a piece of electrical tape and wrap it around where the striker strip would be on the inside. This should create a funnel shape, closing off near the striker strip.

6. Now, take your matches, fold the bundle into thirds, and secure it with a piece of tape about a quarter inch down from the match heads.

7. Next put the matches in the cardboard matchbook funnel you made earlier. You want to push it down so the match heads are just above the striker strip, and the bottom of the matches should stick out slightly.

LIGHTING OFF

1. Next you are going to want to make a pull tab. To do this you are going to fold the top part of the cardboard matchbook over so it's sitting over the tape. Then use another piece of tape to hold it in place. Now you have a pyrotechnic whistle!

2. To ignite your matches, place your finger in the ring you created, hold the bottom of the matches sticking out, and pull. Just like a grenade!

IT'S GETTING HOT! When you pull the two apart, the friction from the striker strip on the inside ignites all the matches at once.

> **PRO TIP:** It may take a lot of force to pull the matches if you wrapped the casing too tight, but if it's too loose the matches will not light at all.

ADDING A FUSE (OPTIONAL)

1. Now to create the fuse. You can use a kind of fuse called visco, which is a safety firework fuse you can get from a firework supply company.

2. It takes about a second to start it off. The only thing you have to do to start the fuse is to knot the end of your fuse and place it in the center of your matches as you roll them.

PYRO ART! As you roll it up you will form a nice little pyrotechnic bouquet of match heads.

3. Make your striker casing like you did before, but this time take your fuse and push it through first.

4. Wrap the loop into position and your pull-string fuse igniter is complete!

Put on your cool safety goggles and experiment. See what happens when you pull the loop fast or slow. Either way, you'll get a bright fire and a shower of sparks!

Ever wanted to make foam flying balloons?
Here's the device that will make that happen!

HELIUM CLOUD GENERATOR

37

SAFETY KEY:
+ Hot objects + Sharp objects

SKILL LEVEL:
EASY
INTERMEDIATE
ADVANCED

APPROXIMATE TIME:
45 minutes

WHAT YOU'LL NEED:
+ Straight-walled pan
+ 6 feet of thin vinyl tubing
+ Tank of helium
+ Bubble soap
+ Hot glue gun

+ Lighter/torch
+ Pins

LET'S BEGIN

SETUP

1. Using your hot glue gun, plug up one end of your vinyl tube.

2. Then glue the tube to the center of the pan. Add some glue over the top to secure it in place.

3. In a spiral shape starting at the center of the pan, add hot glue and stick the tubing to the pan. Using extra glue, secure the tubing near the edge and in at least one more spot in the middle of the coil.

1. Now take your lighter and heat up the end of your pin.

2. Then use the hot tip of your pin to poke holes in the vinyl tubing. Poke holes about every three quarters of an inch around the tube.

3. Now you want to test that your holes work. The easiest way to do that is to blow water down through the tube and it will create a tiny fountain so you can see which holes may have become plugged back up.

MAKING OUR BUBBLE SOAP

1. Regular bubble solution is too heavy, so we will want to water it down. For your bubble solution you will want one part bubble soap and seven parts water.

2. Pour your new bubble solution into the tin and gently stir so that you have a nice, even solution. You want enough solution to cover the tubing but still leave edges of the pan exposed.

FINISHING OUR FOAM BUBBLER!

1. The last step is to attach the other end of your vinyl tubing to the nozzle on your helium tank.

2. All you have to do is press down on the helium spout for the helium to start flowing!

PRO TIP: To make better bubbles let the helium out of the valve slower.

No explosions, just pure fun with bubbles! You'll feel like an evil scientist as you watch your bubble creature grow taller and taller until it finally flies away. Chase after it, cut it, chop it up, or play with it any way that you like.

FUN FACT: The helium we buy in cylinders is produced by the natural decay of radioactive elements—principally thorium and uranium—in the earth's crust.

Set off a rocket with just the touch of a button!
And get this—you might not even have to head to the
store. Start igniting with what you already have
lying around the house. Detonation from the comfort
of your own home!

ROCKET IGNITERS

38

SAFETY KEY:
+ Fire

SKILL LEVEL:
EASY
[INTERMEDIATE]
ADVANCED

APPROXIMATE TIME:
35 minutes

WHAT YOU'LL NEED:
+ Old phone charger
+ Paper matches (or regular
  wooden matches)
+ Electrical tape
+ Scissors
+ Alligator clips

LET'S BEGIN

SETUP

1. First cut off the head of the charger and then cut the cable into smaller pieces about two inches long.

CREATING THE WIRES

1. If you remove the outer jacket of the wire, you will see that there are two insulated wires inside, made of very thin-stranded copper wires.

PRO TIP: For this project, the thinner the wires are, the better the igniter will work.

2. You will not need the outer layer, so you can pull that off.

3. Now you can strip the ends down about half an inch, revealing the thin copper strands.

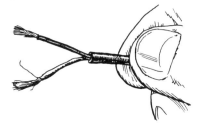

4. Carefully single out one copper strand from the bundle of wires and pull it off to the side.

5. You don't need the other strands, so you can twist them together and cut them off at the base.

6. If your other end has nylon cordage mixed with the copper wires you will want to remove that.

> **PRO TIP:** The best way to eliminate the nylon cordage is to lick it with a flame from a candle or a barbecue igniter.

7. Now hold the two cables side by side about a quarter inch apart and twist the bridge wire and single copper wire with the other cable. Twist it around so it meshes with the other strands. This prevents the bridge wire from unraveling.

8. To hold the wires in place, place them onto the center of a piece of tape.

ATTENTION TO DETAIL: Make sure the bridge wire is about a quarter inch above the tape and then fold the other strand into the tape.

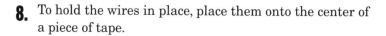

THE MATCH

1. Now you will want to modify the match heads. Take the matches one at a time and carefully slide the match head down the edge of a pair of sharp scissors. You will want to make a small groove down the center of the tip. Using our wires attached to the tape, attach the match so the groove and bridge wire are lined up.

2. Then secure it firmly in place by wrapping one side of the tape tightly over one end and taking the other end of the tape and securing it in the other direction.

LIGHTING IT OFF

1. For one finishing touch, it's good practice to burn the nylon strands on the igniter leads as well.

2. Then twist them together tightly so they'll make better contact.

READY TO GO TO WORK!

1. To test them out, run alligator clips from the ends of the wires to two nine-volt batteries, in parallel.

2. All it takes for ignition is a touch of the wires to the battery terminal.

AUTOIGNITION! It all happens very quickly. The matches light off because when the circuit is completed, over six amps of electrical current surge through the tiny bridge wire on the top. The electricity gets the wire so hot that it ignites the chemicals in the match head, causing it to burst into flames.

There's more: If you want to challenge yourself a bit more, grab an old game console controller (like one from a Nintendo 64) and convert it into a rocket-launch controller!

For practical use or just for fun! This device made with PVC pipe and a clothes hanger will launch about thirty feet.

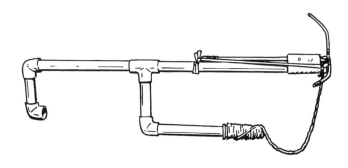

COAT HANGER GRAPPLING GUN

39

SAFETY KEY:
+ Sharp objects

SKILL LEVEL:
EASY
INTERMEDIATE
ADVANCED

APPROXIMATE TIME:
30 minutes

WHAT YOU'LL NEED:
+ $\frac{1}{2}$-inch and $\frac{3}{4}$-inch PVC pipe
+ Wire clothes hanger
+ Nylon string
+ Rubber surgical tubing
+ Wooden dowel
+ Electrical tape or duct tape
+ Baling wire
+ Wire cutters
+ PVC glue
+ Plastic cone from an empty spool of thread

LET'S BEGIN

BUILDING YOUR GRAPPLING HOOK

1. Cut your wooden dowel to about one foot in length. Place aside.

2. Straighten out your wire hanger. Using the wire cutters, clip off the top hanger just below the metal twist. Now straighten out the wire into a long piece.

SHAPE IT!

1. We're making a three-prong hook. First, bend about a six-inch, 45-degree reserve on one end. Then start bending the remaining wire into a hook, bending out and in until you have three spikes.

2. Take the remaining wire and bend it down to meet with the first reserve.

3. Wrap rubber baling wire around the structure to shore up the piece. Once that is done, bend the wire to create your hooks.

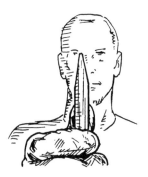

4. Take the wooden dowel and sand it down to get a tapered point.

5. The wedge will fit between the two lower bent wires. Secure with more baling wire, then wrap in electrical tape.

BUILDING YOUR LAUNCHER

1. Cut three inches of ½-inch PVC pipe.

2. Cut eighteen inches of ¾-inch PVC pipe.

3. The ½-inch pipe is going to sit on one end of the ¾-inch pipe. To make a flat, smooth connection, sand down one side of each pipe so they sit flat against each other.

4. Use PVC glue on the smooth surfaces and place together. To secure the pieces you may also want to use duct tape.

5. Cut one foot of surgical tubing (this will be used to launch the grappling hook).

6. Loop the tubing behind the ½-inch support piece and attach the two open ends to the ¾-inch PVC pipe with a zip tie.

7. For the trigger mechanism, cut a hook from a piece of wood. Drill a small hole in the middle of the piece. Drill a hole about four inches from the other end of the PVC pipe. Thread wire from the hanger through the trigger hole and the PVC pipe so the trigger rests on top of the pipe. Twist the wire to secure.

8. Cut into the end of the grappling hook so the trigger has something to grab on to.

9. Build out the launcher so it rests comfortably to your body. Use a T-connector on one end and cut another five-inch piece and insert it onto the bottom. Add an elbow piece.

10. Cut another ten-inch piece and insert that onto the elbow piece.

11. Obtain a hollow plastic cone from a spool of thread. Place it on the end of the ten-inch piece. Drill a hole in the cone and the PVC pipe and use a dulled nail to connect the two.

12. Build out the rest of the gun with another ten-inch pipe secured in the other T-connector end followed by another elbow piece and a short five-inch piece coming down from there, and rounded out by another elbow piece.

13. Measure out thirty-five feet of nylon string. Attach one end with tape to the grappling hook.

ATTENTION TO DETAIL: Make sure not to cover the trigger notch!

14. Drill another hole in the plastic cone, thread the other end of the string through it, and knot it off.

15. Wrap all the string around the cone.

16. Feed the hook through the ½-inch pipe, attach it to the trigger, release the trigger, and watch the hook soar.

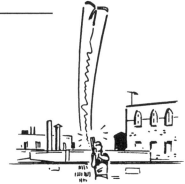

PRO TIP: Adjust your rubber tubing to get the best possible launch!

FUN FACT: Grappling hooks are a tool used by combat engineers. After launching the hook forward they set off trip wires by pulling the hook back.

It may not be small enough to fit on a small table, but it is *definitely* a noteworthy weapon to add to your arsenal of homemade weaponry!

A liquid that is attracted to magnets? Sounds bizarre ...which is the perfect reason to get started!

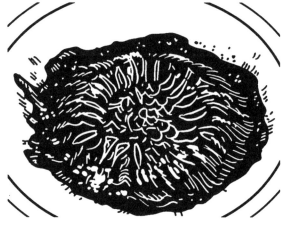

FERRO FLUID

40

SAFETY KEY:
+ Harmful liquids

SKILL LEVEL:
EASY
INTERMEDIATE
ADVANCED

APPROXIMATE TIME:
15 minutes

WHAT YOU'LL NEED:
+ A plastic cup
+ A plate
+ Synthetic black iron oxide
+ Magnetic motor oil
+ Magnetic

LET'S BEGIN

1. Pour several ounces of motor oil into a plastic cup.

2. Add an equal amount of black oxide and stir until it becomes thick.

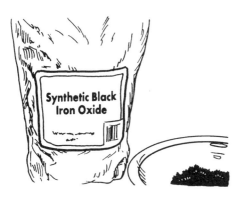

3. Pour the liquid onto a plate.

4. Place magnet under plate.

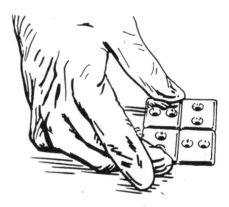

5. The liquid should react and shapes should start appearing.

WHAT IS THIS THING?! The basic idea is it is a liquid that's attracted to a magnetic field. Ferro fluid needs three main components: a magnetic powder that can be suspended, a fluid to suspend it in that will stick to the microscopic particles, and a surfactant (surface-active agent). All three elements are necessary for your ferro fluid to give you the best results!

PRO TIP: If the black oxide begins to separate from the oil, thicken the base with several more plates and then place the magnet under it again.

TEST IT OUT: Now the fun part! Grab some magnets and see what happens. Different magnet strengths should result in different liquid shapes and sizes. If the magnet is too powerful, it may just pull the magnet powder right out of the oil, but try it anyway! Play around with your substance. Move it around, make it dance, or even see what happens when you freeze it with liquid nitrogen. This stuff may be a little creepy... but it's definitely cool.

Don't waste them all on s'mores and hot chocolate! Marshmallows are the new upgraded artillery.

MARSHMALLOW GUN

41

SAFETY KEY:
+ Saw or sharp blade

SKILL LEVEL:
EASY
INTERMEDIATE
ADVANCED

APPROXIMATE TIME:
45 minutes

WHAT YOU'LL NEED:
+ Long piece of ½-inch PVC pipe and assorted ½-inch T-connectors and elbows
+ Mini marshmallows
+ Hacksaw

LET'S BEGIN

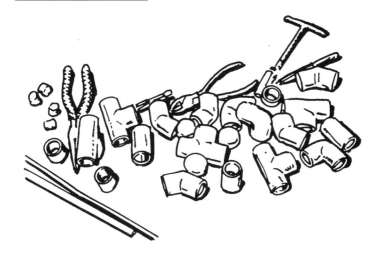

1. Cut your PVC pipe into several lengths: one eight-inch piece, three five-inch pieces, two three-inch pieces.

> **PRO TIP:** To cut your PVC pipe, you can use any saw (like a hacksaw). For best results, get your hands on a PVC cutter specifically designed for the job.

2. Using a T-connector, connect the eight-inch pipe to one of the five-inch pipes.

> **PRO TIP:** As you connect your pipes, make sure they are clean and completely smooth on the inside since the marshmallows are gravity-fed and need to slip through easily to the gun.

3. Use another T-connector to connect the five-inch pipe to a three-inch pipe.

4. Connect the other two five-inch pipes to the open bottom of the T-connectors.

5. Add an elbow to the three-inch pipe and place the other three-inch pipe onto that.

6. Connect another elbow to the last three-inch pipe.

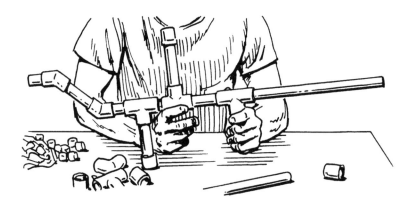

ALSO TRY: Use a tapered drill bit to widen your pipes and ensure a large enough opening.

Ready to fire! Place mini marshmallow in the elbow opening and blow. The marshmallow should shoot out.

PRO TIP: Something you might notice is that there is no standardized size of marshmallow. Use the smaller ones that can go through any twist or turn of the pipes.

IT DOESN'T END HERE! A marshmallow gun is a device that can always be upgraded and taken to the next level. Get creative and design your pipes to give you the sleekest look and the best shooting effect.

SECTION THREE

ADVANCED PROJECTS

If you want to upgrade your marshmallow gun to the next level, here's how to turn it into a semiautomatic.

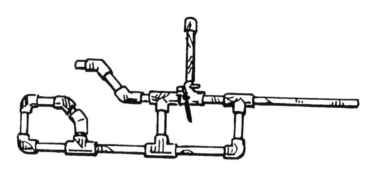

SEMIAUTOMATIC MARSHMALLOW BLASTER

42

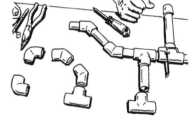

SKILL LEVEL:

EASY

INTERMEDIATE

ADVANCED

APPROXIMATE TIME:

3 hours

WHAT YOU'LL NEED:

+ Marshmallow gun (page 194)
+ PVC pipe cutter
+ Two 45-degree PVC pipe pieces
+ 1 PVC pipe cap
+ 1 T-shaped PVC piece

+ 3/8-inch wooden dowel
+ 3/8-inch drill bit
+ Wire hanger
+ 4 small screws (Phillips pan head #6 × 3/8 in.)
+ Sandpaper

PRO TIP: If you want to make a shoulder rest as well, you'll need to get a total of two 45-degree connectors, two T-shaped connectors, and two 90-degree elbows with spare PVC pipe in between, based on your comfort.

LET'S BEGIN

ORIGINAL MARSHMALLOW GUN MODIFICATION

1. Modify the original blow hole by swapping out the 90-degree pipe pieces for two 45-degree pipe pieces.

MAKE THE MAGAZINE

1. Cut the body of the gun in half.

2. Cut and insert PVC pipe at the top of the T-shaped PVC piece.

3. Connect the body of the gun to the T-shaped piece and add the cap to the top of the new PVC pipe magazine.

PRO TIP: Use a tapered drill bit to widen the hole of the PVC pipe magazine to help make the marshmallows shoot more smoothly.

1. Using the ⅜-inch drill bit (that's the same size as your wooden dowel), drill into the T-shaped piece while it is connected to the magazine. Make sure the holes align.

2. Cut the wooden dowel so that it is 1½ inches long.

3. Round the dowel down using sandpaper to turn it into a wooden peg.

4. Insert the peg into the hole you just made in the magazine just enough to keep the marshmallows from falling down the pipe.

MAKE THE MECHANISM

1. Using a drill bit that is the same size as the wire hanger, drill a hole through the wooden peg.

PRO TIP: When inserted, the hanger wire should be loose enough that the peg can spin, but not so loose that there is a lot of "play."

2. Leaving about three inches of wire on one side, bend the piece of hanger so it goes down below the PVC pipe, extends underneath by a couple of inches, and comes back up on the other side.

3. Bend the wire so that it creates square angles that are flush against the side of the pipe.

4. Now that the wire has come back around, bend the wire so it can go back through the dowel. The bent wire should be a couple of millimeters away from the first wire.

5. Drill another hole in the peg that is a few millimeters away from the first hole. Insert the wire through the dowel again. Clip the extra wire, but leave the three inches of extra wire on the one side.

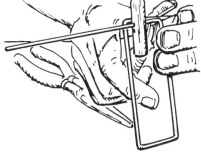

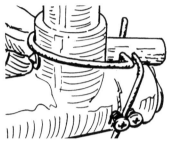

1. Create a total of four holes with the drill bit (two holes on each side) for the small screws to be inserted into. The two holes should be on either side of the wire so that the wire is in between the two screws. Do this to both sides.

2. Using the extra three inches of the hanger wire, push back to create a trigger. Test the new trigger to make sure it can pull the peg all the way back.

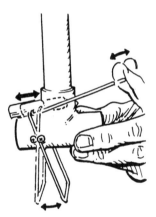

3. While the trigger is pulled back, bend the three inches of wire around the PVC pipe into a hook shape.

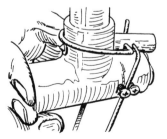

| FINISH |

Now that you've created the semiautomatic portion of the gun, insert the mechanism back into the gun. You'll now have a semiautomatic gun that can shoot marshmallows.

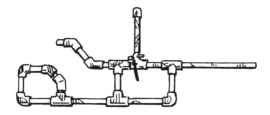

PRO TIP: Create a shoulder rest by using two 45-degree pipe pieces and two 90-degree elbow pieces.

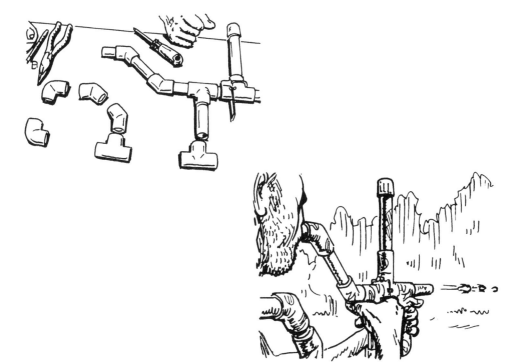

Whoever said being a superhero was easy? The first step is to make yourself some cool tech. Here's how to turn some old shop accessories into high-speed superhero throwing stars!

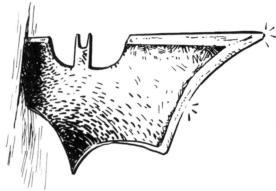

SUPERHERO SHURIKEN (DIY BATARANG)

43

SAFETY KEY:
+ Adult supervision + Safety glasses + Power tools
+ Sharp objects + Check legalities (may not be legal to own or make in many places)

SKILL LEVEL:
EASY

INTERMEDIATE

ADVANCED

APPROXIMATE TIME:
2 hours

WHAT YOU'LL NEED:
+ Scrap metal (10-inch circular saw blade works fine)
+ Printed image of the batarang
+ Glue stick
+ Clamps (or bench vise)
+ Jigsaw with metal-cutting blade
+ Metal file set
+ Sanding sponge, or fine-grit sandpaper

OPTIONAL:
+ Spray paint suitable for metal

LET'S BEGIN

ROUGH OUT THE BASIC SHAPE

1. Glue your paper template to the surface of the metal. Smooth out any creases and let it dry for about ten minutes.

2. Clamp the metal securely to your work surface, preferably on top of some scrap wood to protect it.

3. Now it's time to cut the metal. Start by cutting a straight line across the top of one wing first, then continue this line across both wings. This will make cutting out the head, and intricate bat ears, a little easier.

PRO TIP: Be sure to cut very slowly, and use oil to help dissipate the heat. The tight curves require a steady hand, and cutting too fast will ruin your blade within minutes.

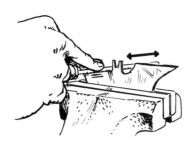

1. At this point you should have something that resembles your throwing star template, but in metal. Don't worry if it's rough around the edges. That's perfectly normal, and just needs to be cleaned up a bit. Secure it with a clamp or bench vise and use the flat side of your file to grind down the jagged points along the sides, and use the rounded edge to smooth out the curved areas.

HELPFUL HINT: Try filing at an angle to add a subtle bevel to the wings. This makes the wing tips look sharper, and worthy of your caped crusader utility belt.

2. It looks awesome in silver, but for a professional look, go one step further and paint it black. Push a wing tip into a piece of cardboard so it stands upright, and hit it with a few shots of spray paint.

PRO TIP: Some spray paints, like Krylon, can cure in as little as ten to fifteen minutes and are great if you're impatient or short on time.

3. Rub sandpaper or a sanding sponge on the beveled edges to re-expose the steel and give your throwing star that cool two-toned look.

GOTHAM READY! Your Superhero Shuriken is finished and all set for testing.

TARGET PRACTICE

Walk three paces from your target, hold the throwing star at the bottom, and throw it with a quick flick of the wrist. With a bit of practice it'll complete one full rotation and penetrate the target with a satisfying thud.

> **HELPFUL HINT**: Adjust your distance in small increments to find the sweet spot where it works best for you. A distance of three steps is roughly one full rotation, so from six paces it should spin twice.

You could also use a twelve-inch two-by-eight board to build yourself a throwing post, and attach a dartboard to the center to hone your skills even further.

> ☠ **WARNING**:
> It's important to treat your Superhero Shuriken with the same respect as a throwing star or a knife. Make sure you've got an adult sidekick nearby, and avoid throwing close to people, animals, or other people's property.

Who would've guessed a rusty old table saw blade could make such a sleek batarang? Whether you're chasing down the Joker or simply practicing on a DIY target, you'll be sure to slay like a superhero!

> **FUN FACT**: Known for being Batman's customized throwing star, the batarang was also used by Batgirl. Even Robin (back when he was called "Nightwing") had his own version of the batarang, which he gave the not-so-intimidating name "Wing-Ding."

Poof goes the cannon! Whether you call it a Smoke Ring Shooter, Air Vortex Cannon, or Airzooka, this awesome toy can shoot down solid targets, suck up and fire out dry ice, and launch giant rainbow air rings over thirty feet.

SMOKE RING SHOOTER

SAFETY KEY:
+ Sharp objects

SKILL LEVEL:

EASY

INTERMEDIATE

ADVANCED

APPROXIMATE TIME:
2 hours

WHAT YOU'LL NEED:
+ Plastic bucket (without a spout around the lid)
+ Wood stake
+ Drill
+ Saw
+ Three #12 24 × 1-inch machine screws with matching nuts
+ Clear shower curtain
+ Clear packaging tape
+ Nylon screw and nut
+ Bungee cord (with bead on end)
+ Spray paint
+ Empty soda bottle

44

LET'S BEGIN

PREPARING THE BUCKET

1. Use a utility knife to cut a circle out of the bottom of the bucket, leaving a 1½-inch rim.

2. If your bucket already has a string handle, cut it off now. Line the wooden stake up on the bucket and measure how long you want the support for your custom handle to be. Cut the stake to size.

3. Place your support rail flush against the bucket. Using a Sharpie, mark three points near the top, middle, and bottom. Drill through the wood and the bucket in each of the three spots.

4. Hold the remaining piece of the stake against the half you have already cut and drilled. Measure how long you want it to be as a handle. Cut this piece.

5. Drill two pilot holes through the support rail into the handle. Securely fasten both pieces with screws.

PRO TIP: Sand your handle down to make it more comfortable to hold on to.

6. Cut the top two inches off your empty soda bottle. Hold a nail by the tip with a pair of pliers and then use a lighter to heat up the head. While it's still hot, melt a small hole near the neck of the bottle. Melt a second hole on the other side.

7. With a Sharpie, mark two points on opposite sides of the bucket, almost at the bottom. Heat up the nail again and melt holes in the bucket where you marked it.

PRO TIP: By melting the hole instead of drilling it, you get extra strength from the melted plastic.

CONSTRUCTING A PLASTIC DIAPHRAGM

1. Spread out your plastic shower curtain. Place the top part of the soda bottle in the center of the sheet and trace around it with your marker. Move the bottle underneath the sheet of plastic in line with the circle you just marked. Screw the bottle cap on from the top of the sheet, securing the bottle to the plastic.

2. Place the bucket in the center of the sheet. Press the soda bottle and the plastic halfway down into the bucket. Mark where the plastic and the top edge of the bucket meet.

3. Remove the plastic and measure the distance from the soda bottle to the mark you just made. Now measure that same distance from the soda bottle all around the plastic in a circle. Connect the marks together to get one continuous circle.

4. Cover the whole circle in a layer of packaging tape. Once the whole circle is covered, cut it out by following the line you just drew, and reattach the soda bottle to where it was before.

5. Drill a hole through the soda cap and plastic.

PRO TIP: If your hardware store doesn't carry nylon screws, try toilet seat–hinge hardware. A metal nut and bolt would also work.

6. Cut your bungee cord to remove the plastic ball, setting the cord aside for later.

7. Thread the nylon bolt through the plastic ball and then through the hole in the soda cap. From the inside of the bottle top, thread the nylon nut onto the bolt.

Before you assemble all your pieces, spray paint your Air Vortex Cannon to match your personal style, so long as you don't paint the plastic sheet.

PUTTING IT TOGETHER

1. Center the bottle top inside the bucket. Attach the edge of the plastic sheet to the edge of the bucket using a piece of duct tape.

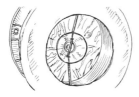

2. Take a second piece and tape it directly across from the first. The next two pieces are going to be at a 90-degree angle from those. In between those four corners, make an accordion fold with the plastic to get it to lie flat against the bucket. Tape over any gaps so that the whole edge of that plastic is secured to the bucket.

3. Now all that is left is to attach the handle and make it spring-loaded. Thread your bungee cord into one of the holes in the inside of the bucket, through both of the holes in the soda bottle top, and then out through the other bucket hole.

4. Mark the bungee cord where it leaves the bucket. Now stretch it out and tie an overhand knot.

5. Place the handle and support rail along the bucket aligned with the drilled holes. On the inside of the bucket, attach three nuts and then tighten with a screwdriver.

SHOOTING YOUR AIR VORTEX CANNON

Simply point the open end of the bucket at your target, pull back the plastic, and let go! As the air is forced out the front of the cannon, the air in the middle moves faster than the air on the sides, forming a ring. Try setting up a pyramid of paper cups to knock over! Or experiment with dry ice.

Launch it at targets, shoot it long distances, or just have fun messing with your friends, family, or pets. After all, it's just air. You can't go wrong with this cool contraption!

FUN FACT: These cannons can come in a variety of sizes. The largest air cannon made was built in the Czech Republic and shot down a wall of cardboard boxes from three hundred feet away.

Looking for something crafty and stealthily artistic? For this project, use sand and plaster to make a simple backyard foundry that can melt scrap metal in seconds, but is still a pleasant decoration. No one will ever know that the houseplant is actually a homemade furnace!

MINI METAL FOUNDRY

SAFETY KEY:
+ Eye protection

🔥 FIRE WARNING:
+ The mini foundry gets so hot on the inside it will melt soda cans within seconds. Use extreme caution.

SKILL LEVEL:

EASY

INTERMEDIATE

ADVANCED

APPROXIMATE TIME:
4 hours

WHAT YOU'LL NEED:
+ 25-pound bag of play sand
+ 25-pound bag of plaster of paris
+ Buckets
  2.5-quart plastic bucket
  5-quart big-mouth plastic bucket
  10-quart steel bucket

+ Steel wool
+ Plastic tablecloth
+ Hacksaw and vise
+ 1 × 12-inch steel pipe
+ 1-inch PVC coupling (Slip x FIPT)
+ 1-inch PVC pipe
+ Two 4-inch U-bolts

+ Spray paint
+ Charcoal briquettes
+ Channel locks
+ 1 3/8-inch hole-cutting saw
+ 3-inch hole-cutting saw
+ Hair dryer
+ Old fire extinguisher or clay graphite crucible

LET'S BEGIN

MAKING THE MIXTURE

1. First prepare your steel bucket by lining it with steel wool.

2. You can quickly combine the ingredients by using the 2½-quart plastic bucket as a measuring tool to combine ¾ bucket (21 cups) of plaster of paris, 1¾ buckets (21 cups) of sand, and 1¼ buckets (15 cups) water. From the moment the water touches the dry mix you'll have about fifteen minutes before it all hardens, so immediately use your hands to mix it all up.

3. Transfer the plaster of paris mixture to the steel bucket you previously lined with steel wool. Pour as slowly as is practical to minimize splattering. The mixture should reach three inches from the top.

4. Use the plastic measuring bucket to form the center of the foundry. Fill with water and push the weighted bucket into the center of the steel bucket. The mixture should rise upward but not spill out. Hold down for two to three minutes to set.

5. Using a paper towel and some water, clean up the edges around the bucket. Let sit for about an hour.

6. Dump the water from the bucket. Use pliers or a pair of channel locks to gently pull one edge toward the center. Now grip the channel locks with both hands and give it a twist. The whole bucket should pop loose leaving a smooth hole in the center of the steel bucket.

INSTALLING AN AIR SUPPORT AND LID

1. Center your 1⅜-inch (35mm) hole-cutting saw with the top raised ring circling the steel bucket. Once you cut through the metal, burrow down at about a 30-degree angle, which should be fairly easy as the plaster has not fully cured yet.

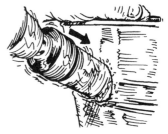

PRO TIP: We want the hole to be tight and at a downward angle so if the crucible fails, the molten metal will stay in the foundry instead of spilling toward you.

2. Make your blower tube by connecting your one-inch steel pipe to the one-inch PVC using the coupling screw. Twist the threaded side onto the steel pipe and push the PVC through the other end with the slip adapter. Remember that the steel portion is what is inserted into the foundry.

BUILDING YOUR FOUNDRY LID

This will serve as your vent as well as allow you to melt metal without removing the lid.

1. Fill the big-mouth bucket with cups plaster of paris, 10 cups play sand, and 7 cups water. Mix well.

2. Stand two four-inch U-bolts upright in the mixture. You can place a three-inch-wide can in the center of the mixture to cast a vent hole at the start of the process. Or, when the mixture is set, use a three-inch cutting saw to drill a hole right in the center.

3. Wait an hour for the plaster to set and then pop the lid free from the bucket.

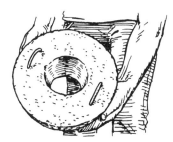

MAKING THE CRUCIBLE

HEATING UP! A fire extinguisher doubles nicely as a crucible to prove this project could be completed in an apocalyptic situation, but it's not the only option. You can buy an excellent clay graphite crucible online for around $30.

1. If you decide to use an old fire extinguisher, note that if a magnet sticks to the side, then it's steel. If it doesn't, then it's aluminum and it won't work for you.

2. Unscrew the valve from the top of the extinguisher to depressurize. Place the tank in a vise and then cut it in half with the hacksaw. The bottom part is now a steel cup three inches in diameter and five inches tall.

USING THE HOMEMADE FOUNDRY

1. Use duct tape to attach a hair dryer to the end of the blower tube. This will supply air to the flames.

2. Toss charcoal briquettes into your foundry, light, and put the lid on.

WARNING:
☠ Charcoal used for barbecuing is specifically designed to make the coals burn slower at a lower temperature, so to achieve a hotter temperature use lump charcoal, which is basically just pyrolyzed wood. Lump charcoal will emit a shower of sparks, so wear long sleeves and gloves. Also use a lower setting on your hair dryer as you don't need to pump that much air into it to get the same temperatures.

3. Your foundry will soon reach temperatures around 2,000 degrees Fahrenheit, which is hot enough to melt not just aluminum, but copper, silver, and gold. Try melting soda cans and scrap metal. You can also use your foundry as a blacksmith forge, or even to roast hot dogs for a super-charged cookout—it is powered by charcoal after all.

PROTIP: As soon as you turn the hair dryer off, pull the tube out of the metal foundry. If you leave the tube in the foundry with no air blowing, the heat will travel up and melt your hair dryer.

Now you have a handmade mini metal foundry that can reach unbelievable temperatures. And when it's not in use, drop a potted plant inside and instantly transform it into fashionable home decor. It's important to remember that this is an entry-level system. If you want to get serious, I recommend investing in proper refractory materials. Let the broiling begin!

FUN FACT: If you could take a time machine back to Mesopotamia in 3200 BC, you would find the first metal castings. Fast-forward to today and metal casting has become a $33 billion industry in the US, with 90 percent of all manufactured products containing some metal-cast part.

It appears to be so simple . . . puzzling, is it not? Well, that's exactly what it is! This paint-stick puzzle box may look elementary, but there is much more to it than meets the eye!

PUZZLE BOX

46

SAFETY KEY:
+ Sharp objects

SKILL LEVEL:

EASY

INTERMEDIATE

ADVANCED

APPROXIMATE TIME:

2 hours

WHAT YOU'LL NEED:

+ 2 packs of paint sticks
  (10 sticks per pack)
+ Hot glue
+ Toothpick
+ Small dowel
+ 2 small magnets
+ 1 nail and small screw

+ Red chestnut wood
  stain
+ Polycrylic finish
+ Saw
+ Sandpaper
+ Sledgehammer
+ Hacksaw

LET'S BEGIN

CUTTING YOUR PAINT STICKS

1. Measure, cut, and sand your paint sticks to the following sizes:

+ **Five 8$\frac{1}{2}$ inches**
+ **Two 7$\frac{1}{2}$ inches**
+ **Two 6$\frac{1}{2}$ inches**
+ **One 5$\frac{1}{2}$ inches**
+ **Eight 3$\frac{3}{4}$ inches**

+ **Four 3 inches**
+ **Five 2$\frac{1}{2}$ inches**
+ **Nine 1$\frac{1}{2}$ inches**
+ **One 1$\frac{1}{8}$ inches**
+ **One 1 inch**

BUILDING A LOCKING PIN

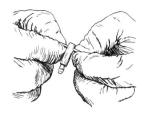

1. Pound the head of the nail with your sledgehammer until it's flat.

PRO TIP: A belt sander will make smoothing down your cut pieces easier and more professional. You can also grind the hammered nail tip to resemble an improvised flathead screwdriver.

2. Using a hacksaw, cut half an inch from the top of the nail.

3. Choose a drill bit that is a comparable size to the nail itself. Secure your dowel upright in a vise and drill half an inch straight down into the center.

4. Use hot glue or superglue to secure your modified screwdriver into the dowel itself, flattened end sticking out.

5. Choose a drill bit that is the same size as your toothpick and drill a hole through the dowel, about an inch from the end. Push the toothpick through the hole to about the middle. Glue the toothpick into the hole and then gently break off the ends and smooth those over with sandpaper.

ASSEMBLING THE BOX

You will need the following paint sticks:

+ **Four 8$\frac{1}{2}$ inches**
+ **Two 7$\frac{1}{2}$ inches**

1. Lay your paint sticks on a piece of paper. You should have two full-sized sticks on the right side and the 7½-inch stick on the left side.

2. Secure the three sticks together by laying a bead of hot glue down the side of one, and then firmly pressing the edge of another against it. Repeat to have two sets of three sticks. After they dry, sand all the edges. Flip each of them over so the smooth side is facing up, mark-free. These will be the lid and bottom plate of the box.

3. You will now need the eight pieces that were cut to 3¾ inches to be the strappings. Leaving about half an inch at the bottom, lay the pieces horizontally across the lid and bottom plate. The remaining paint stick ends will serve as spacers, so make sure you only glue down every other stick. Wait for the glue to dry and remove the spacers. You should have four wooden straps assembled on the lid and bottom plate.

4. Choose your least favorite plate and set it down with the wood straps on the tabletop. This will be the bottom.

5. You'll now use the last 8½-inch piece to build one sidewall. Bead a line of hot glue along the side of the bottom plate and connect it to the edge of the base.

6. Now grab a three-inch piece, but before you glue it down, you'll want to measure. When you press it up against the sidewall, you want to see a gap on the other side, at least the width of a paint stick for the other wall. When you glue this piece, you'll want to glue the bottom and side touching the long wall.

BUILDING YOUR SECRET COMPARTMENTS

1. Start with a small compartment tucked in the back of the box. Glue three 2½-inch pieces together, using one as a base and the other two as sidewalls, glued to the outside edges of the one you're using as the base. Glue the 1⅜-inch piece to serve as the back of the compartment. This will now fit perfectly at the bottom of the base plate, next to the back wall. There should be a small gap at the back for your magnets—now is a great time to check that they fit!

2. Place your small compartment flush with the back wall. Glue a three-inch piece to the front side of the compartment, off center, to begin building the other sidewall.

3. The second, larger secret compartment requires: two 6½-inch pieces to create the sidewalls; two 3-inch pieces for front and back; two 2½-inch pieces underneath for support. Create a simple rectangle by gluing the two 6½-inch sidewall pieces to the front and back pieces.

---

4. Cut a piece of thin cardboard, like a cereal box, to serve as the bottom.

PRO TIP: Make sure you glue the sidewalls to the *inside* of the front and back pieces, as well as sand thoroughly, so it can easily slide in and out of the box.

---

5. Glue the 2½-inch pieces flat down and against the inside walls for support.

FINISHING UP

---

1. Glue down a 5½-inch piece to complete the box's last sidewall.

---

2. Glue one magnet to the back of the small drawer, and the other to the box on the wall just opposite. Use two *opposing* magnets so the drawer will pop out like a spring when the catch is released.

PRO TIP: If your gap in the back is too wide, just add one more magnet.

3. Glue the lid to the completed base, the strappings facing up. It's important to note there is a notch missing. So hot-glue everything but where that missing piece is, or you'll make a sticky mess. After attaching the lid, it may have arched a little or there may be some excess glue. Slide your compartments back in to make sure they still fit properly.

4. Use your two one-inch square pieces to fill in the holes in the lid and bottom plate. Cut a 45-degree angle off of the square pieces you just attached to the drawers so it won't get caught on the edges when sliding in and out.

5. Now you will need six of the 1½-inch pieces. Glue these to the sides of the box, connecting the strappings on the top and the bottom. Make sure to leave two spaces empty for strategically placed pieces. These should be the middle left on each side.

6. To make the mechanism that allows the box to open, close the small drawer tightly and drill a pilot hole through the center back wall. The screw will bore into the side of your small box, locking it in place. Lock the drawer in place by twisting in a small screw through the pilot hole.

7. Use a ³⁄₁₆-inch bit to drill a hole through the side of the small compartment, near the seam line where it meets the large drawer so it can be covered by your strategically placed strapping. The long dowel should fit snugly inside. It will be most accurate if you drill the hole while everything is locked in place.

8. Flip the box to the other side. The missing strap panel will be the second from the left. Cover with your 1½-inch wood piece and drill through the center, penetrating three layers: the 1½-inch panel, box sidewall, and large mystery compartment.

> **PRO TIP:** The holes for the strapping panel and the mystery compartment need to be as tight to the dowel as they can be. The holes going into the box itself should be looser, so take your drill and open them up now.

9. Withdraw the small compartment entirely out of the box. Cut a ½-inch piece of dowel and push it through the drilled hole so the dowel is extending on the inside. Glue the dowel into place. If the  dowel is sticking out of the outside, you can easily sand it down to make it a smooth surface.

10. Glue the 1½-inch panel down, making sure to only affix it to the outside wall of the mystery compartment, not the wall of the box itself.

11. For the other secret side panel, glue the dowel through the hole. In order to hide the fact that there's a dowel, grind the piece down with the belt sander until it is half the thickness that it was originally. Add an identically sized panel, glue it to the top, and then grind that down as well. This will create a veneer effect, making this piece look just like the others.

12. Bring back the mini screwdriver you made earlier. Drill a ⅜-inch hole on the same side as the small compartment, but in the far right gap. Your dowel should fit snugly in

place, and serve as a first clue to whoever is trying to solve your puzzle box.

13. Go the extra mile by sanding the outside smooth and adding a handsome wood stain and varnish.

14. Use a soldering gun (or permanent marker) to write clues on the box.

Solving puzzles is a new, exciting endeavor every time! But constructing your own puzzle box brings it to a whole other level, especially when you get to watch your friends and family try to solve it. Will they be simply perplexed or completely discombobulated? Best of all, this one costs about $3 in materials, and is a perfect gift, weekend project, or geocache.

FUN FACT: Puzzle boxes produced for entertainment first appeared in Victorian England in the nineteenth century.

Have you ever seen T-shirts launched into a crowd at a concert or sporting event and thought, "Forget the shirt, where do I get that cannon?" Follow these steps and you can make one yourself!

T-SHIRT CANNON

47

SAFETY KEY:
+ Flying objects

SKILL LEVEL:
EASY
INTERMEDIATE
ADVANCED

APPROXIMATE TIME:
2 hours

WHAT YOU'LL NEED:

+ ABS fittings
   2-foot length of 3-inch ABS
   2-foot length of 2-inch ABS
   Adapter from 3 inches to 2 inches
   Two 3-inch couplings
   Adapter from 3 inches to 1$\frac{1}{2}$ inches
+ PVC
   1$\frac{1}{2}$- to 1$\frac{3}{4}$-inch slip to threaded adapter

Six $\frac{3}{4}$-inch threaded close nipples
$\frac{3}{4}$-inch coupling
Two $\frac{3}{4}$-inch 90-degree elbows
$\frac{3}{4}$-inch T-connector
End cap
$\frac{3}{4}$-inch ball valve
+ Thread tape
+ Schrader valve
+ Saw

+ Spray paint
+ Proto-Putty (Chapter 9)
+ Bike pump
+ Black ABS cement
+ ABS to PVC green transition cement
+ T-shirts

LET'S BEGIN

PREPARING YOUR MATERIALS

1. Cut your three-inch tube to sixteen inches in length, and the two-inch tube to fourteen inches.

2. Wrap both ends of the ¾-inch threaded close nipples with a few layers of sealing tape to get a tight connection.

3. Drill a ½-inch hole in the top of the end cap. Push the Schrader valve through the hole.

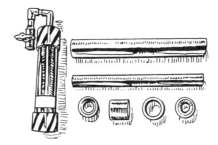

PRO TIP: Instead of a Schrader valve, try a pneumatic quick adapt connector so you can fill your T-shirt cannon with an air hose instead of a bike pump.

4. Spray-paint the ¾-inch ball valve the color of your choosing. Let dry for a few minutes.

1. Screw together your PVC parts. The threaded close nipples connect the elbows, T, end caps, coupling, and ball valve.

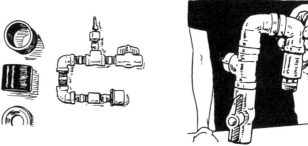

2. Spray-paint the unpainted parts with a second color of your choosing.

PRO TIP: When painting PVC you want to make sure you first do a light dust coat over the whole surface to let the next layer of paint bond to the plastic.

3. Now it's time to assemble the body of the cannon. The sixteen-inch tube goes inside a three-inch coupling, and the fourteen-inch tube slides down inside one of the reducing adapters.

4. Take the 3 to 1½-inch adapter and place this one on the table bevel side up. Put the second three-inch coupling over it. Use black ABS cement to glue the pieces together.

5. Put the smaller tube adapter side down on the table, and put the bigger tube over it. Cement where the coupling and adapter meet.

T-SHIRT CANNON

MAKE YOUR PLUG

1. We'll be making our plug out of Proto-Putty. Mix the Proto-Putty in a paper bowl, and when it's good to go, toss some baby powder on a cutting board. Place the leftover one-inch piece of scrap ABS from the three-inch tubing on the cutting board and pack the Proto-Putty inside the mold.

2. Press down with a large flat object to ensure it dries smoothly. Set for ten minutes.

3. Now using a utility knife, cut the putty around the ring to loosen it, then pop it out. Clean up the edges with scissors and your Proto-Putty plug is complete.

4. Push into the three-inch ABS tubing until it makes contact with the two-inch tube inside. You can now pressurize the system.

FINISHING UP

1. Use ABS cement to secure the last three-inch coupling onto the end of the cannon.

2. Now take the green transition cement to attach the PVC handle to the ABS cannon. Coat the inside of the cannon adapter and the outside of the PVC handle—the opposite end from the ball valve. Quickly insert the PVC handle into the end of the cannon.

3. Now's the time to customize! Use duct or electrical tape to decorate your cannon as a one-of-a-kind creation.

USING YOUR T-SHIRT CANNON

1. Hook up a bike pump to the valve and start filling your cannon with air. The Proto-Putty plug will shift upward to block the interior tube, and the air will fill up the large chamber on the inside. Pressurize up to 60–80 psi.

2. Shove your shirts down the barrel, packing them tightly with a stick.

3. Release the handle. The air will rush out, forcing the plug to slam down to the bottom of the barrel, the remaining air shooting out the center of the tube and launching your T-shirts quickly and powerfully—around forty feet!

ALSO TRY: Why stop at T-shirts? Try shooting Nerf darts, or really anything else you can fit down the barrel.

Shout-out to our friend Ben and his Coaxial Potato Cannon for being the inspiration behind this project! Why not use the same concept and common materials from a local hardware store to construct a fully functional semiautomatic Air-Powered T-Shirt Cannon? Just T-shirts at events? This device can do so much more!

If you were stranded in the middle of nowhere, what would it take to create the building blocks of survival—clay? Dig down into the depths of the earth and see just what it would take!

CLAY FROM DIRT

SKILL LEVEL:

EASY

INTERMEDIATE

ADVANCED

APPROXIMATE TIME:

1 day

48

WHAT YOU'LL NEED:

+ Several buckets

+ Water

+ Dirt

+ Cloth/strainer

+ Towel/cloth

+ Paper towels

LET'S BEGIN

SETUP

1. To gather the materials needed, you can go to a local river or pond and start digging down into the dirt as close to the water as is comfortable for you.

2. You want to find dirt that has an almost plastic-like feel, so when you squish it together it will stick in clumps.

FILTERING OUT THE CLAY

1. Once you have your dirt, the first thing you will want to do is extract all the rocks and twigs. To do this, you will want to filter it through an easy purifying process. All you need is a couple of containers and some water. First, fill your container you have your dirt in with water so it becomes super muddy, like a soup. Feels like soup, but don't eat it if you get hungry!

2. Next you will want to stir the mixture so that all the material separates, and then let it settle for about five seconds.

3. Then transfer the liquid on top into a separate container.

  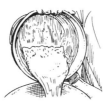

PRO TIP: While you can make clay from dirt in your backyard, it is much better to make clay from dirt near a body of water.

4. Stop pouring when just most of the heavy stuff (gravel, rocks, etc.) is left in the original container. You can now discard the container with the leftover rocks as you will not need any of that.

5. Next you will need to strain your liquid. You can use an old shirt, socks, or just a regular strainer. This process should remove most of the organic material. Before you pour your liquid solution through the strainer into a new container, it is important to stir it again to resuspend the clay particles in the water.

USE YOUR BEST JUDGMENT! Left in your strainer you should see leaves, twigs, and other organic materials. You can repeat this process to filter out any more material you may have missed.

6. You may notice some fine sand left in your container as you pour between buckets. If you keep doing this to clean the bucket each time, you will be left with a very fine clay solution.

LETTING YOUR CLAY SIT AND DRY

Now you'll need to let it settle for a couple of hours. It takes quite some time to let the water filter to the top since the clay is so fine. Here is a way to help speed up the process:

1. First you need to remove the water you have so far from the top (which should have gathered after about an hour).

2. You can pour the water out very carefully, without pouring out any clay material. Or, you can siphon the water out. Using some tubing, you can suck the water, but before it gets to your mouth, pour the water into a bucket and gravity will continue to pull the water out.

3. Now to get even more moisture out of your clay, take a towel or tight-knit cloth; you want it tight enough where the clay will not go through, but the water will.

4. Once it's all poured in, you can tie up the ends for a few hours and let the atmosphere suck most of the water out. After leaving the clay for an hour or so, you may notice that it is still pretty runny.

5. Taking some paper towels, scrape the clay from your towel/cloth and put it in the center of the paper towel and just flatten it out, squishing your clay between two paper towels to remove most of the remaining water.

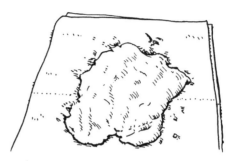

6. You should be able to peel away the paper towels, and have some nice moldable clay. Now that's the best part!

7. Add or extract water as necessary until it reaches a desirable consistency (the thickness is based on preference).

PRO TIP: If you want to save your clay to use later, just wrap it in some Saran wrap.

Now that's an experiment where you can really feel the result. The smooth texture of success! You can carve your clay, roll it, or even fire it in an oven to make pottery, earthenware, or possibly even ceramic! Who knew that river mud could become full-blown clay?

FUN FACT: Clay is a soil that is unlike any other because of its natural plasticity. By baking or firing clay, it goes from being moldable to rigid, resulting in endless opportunities for creativity.

Flaming, colorful, glowing, bubbly vortex fountain? You had me at flaming! This project may not require you to aim and fire, but you'll never want to pull your eyes away from its tornado flames of style.

GIANT FLAMING VORTEX FOUNTAIN

49

SAFETY KEY:
+ Fire

SKILL LEVEL:
EASY

INTERMEDIATE

ADVANCED

APPROXIMATE TIME:
4 hours

WHAT YOU'LL NEED:
+ Bucket
+ 16-inch-diameter plastic tray
+ Drink dispenser
+ Water pump
+ Vinyl hosing
+ Plastic elbow connectors
+ LED lights (optional)
+ Bamboo blinds (optional)
+ Rocks (optional)

LET'S BEGIN

SETUP

1. First you will need a large bucket to act as your cistern down below; a pump that can move up to 550 gallons of water per hour; flexible vinyl housing; a drink dispenser that will house the vortex; and a plastic sixteen-inch-diameter tray.

2. First we are going to strip all the excess pieces from the drink dispenser. You are going to want to remove the spigot, but keep the rubber seal.

3. At a hardware store you can find these plastic pieces that turn at an angle; one end is smaller than the other so they can fit together. The smaller end also fits perfectly with your ⅜-inch vinyl hose.

4. You are going to fit the angled piece through where the spigot was, having the narrow end sticking outside.

5. You will then attach the other half to the first one through the container. You are also going to use your rubber seal to help secure the hole.

6. To make sure this is 100 percent waterproof, you are going to want to hot-glue the rubber seal onto the water container.

DRILLING THE HOLES

1. Next you are going to drill a hole on the bottom of the container in the center with a ⅜-inch drill bit.

2. Then you are going to want to make similarly placed holes in your plastic tray.

NOTE: Any plastic tray about sixteen inches in diameter with a lip around the edges should work.

3. First you are going to drill a hole in the center that is about an inch wide. This is where the water will drain through after it leaves the vortex chamber.

4. Next line the two holes up and place a mark where your next hole should be, which is going to be where the tube will come up to pump water into our tank.

5. Once marked, make a hole using a ¾-inch drill bit.

6. To help your fountain drain faster, you can drill a couple more holes right along the edges of your water container. Make sure they are still inside the bucket or else your

GIANT FLAMING VORTEX FOUNTAIN

water will just drain onto the floor. For these holes you can use a ⅝-inch drill bit.

CONNECTING YOUR VORTEX FOUNTAIN

1. You will only need your hose to run from the bottom of your bucket to the entrance of your fountain, so you can cut your hose to that length. To err on the side of caution cut about six inches longer so you can measure it exactly once it's all attached.

2. Attach your hose to your pump.

> **PRO TIP:** If your vinyl hose is curling too much, you can use some heat to help straighten it out and then let it cool down while holding it straight up.

3. Place your pump with hose attached inside at the bottom of your bucket.

PRO TIP: On your bucket, you will want to cut a notch so the power cord can come out and your tray will still sit flat.

4. Next place your tray on top and run your hose through its designated hole.

5. Place your water container on top of the tray and line the holes up.

6. Now you can mark where to cut your hose to length.

7. Now that everything is lined up and connected, you can add water to your bucket to check to see if it works.

8. You are going to want to make sure you have enough water for this to work. To do so add enough water to cover your pump, then enough water to fill your fountain container, which will ensure that no matter how full your fountain gets, your pump has enough water to keep going.

A water tornado? Now that's awesome!

DESIGNING YOUR VORTEX FOUNTAIN

LIGHT IT UP!

1. If you want to make your vortex fountain colorful, you can do so by adding some LED lights. These lights you can wrap along the bottom of your fountain. To add the lights, you will want to drill a small hole in the bottom of the tray to run the lighting through and then use hot glue to seal it off.

2. When you run your lights up you want to leave some slack so you can still lift the tray on and off.

3. You can use tape to secure your cords down to the bucket.

4. You may notice having the bucket and wires on the bottom is pretty ugly. There are multiple ways you can disguise this. One easy way is by using bamboo blinds. You will want to cut the blinds so they fit just underneath the lip of the tray. You can hot-glue the bamboo together once wrapped all the way around.

ALSO TRY: Add some stones and rocks to your tray for added decor!

THE BEST PART! Want to add some flames to your vortex? Right now there are tiny bubbles swirling around your vortex. Add a hose going into the bottom that wraps around in a circle with some smaller holes pricked in it. Now add some pressure and see if you can add some larger bubbles to the vortex as well.

GUESS WHAT? These aren't oxygen bubbles—they're propane!

So now we have a colorful glowing *fire* vortex fountain. Too cool!

Getting hungry? Grab a few simple tools and build yourself a little machine that cooks a number of tasty snacks!

CARDBOARD BOX DEHYDRATOR

50

SAFETY KEY:
+ Soldering

☠ WARNING:
+ Do not leave the Cardboard Box Dehydrator unattended while in use. Always unplug to avoid fire hazard after use.

SKILL LEVEL:
EASY
INTERMEDIATE
ADVANCED

APPROXIMATE TIME:
45 minutes

WHAT YOU'LL NEED:
+ A cardboard box
+ A lightbulb (250-watt bulb, ideally)
+ A small fan
+ Aluminum foil
+ Strong tape (metal tape works the best)
+ Soldering iron
+ Electrical tape
+ Cooling racks
+ Dowels
+ Heat lamp

LET'S BEGIN

PREPARE YOUR BOX

1. Grab your cardboard box (any size will do, but remember that you will want to place racks inside of it to hold your food).

2. First you'll want to line the inside of your box with aluminum foil. Tape it down shiny-side out, lining both the box and its lid.

SO SHINY! The reflective aluminum foil is going to help the heat bounce around and not get absorbed by the cardboard.

YOUR BIGGEST FAN

1. Now it's time to prepare your fan for installation.

PRO TIP: Sometimes thrift shops will carry old fans. For instance, they could have a fan from an old computer (used power supply) that could be conveniently upcycled for this project. The power supply should indicate that it has an output of around 12 volts DC current.

TEST IT FIRST: Before you assemble further, do a quick test to see if your power supply is working: Connect the wires of your power supply to a wire, then plug it in. If the wires are connected in the right direction, your fan should work!

2. Now solder the wires together to connect them permanently using a soldering iron and electrical tape.

INSTALLING YOUR FAN

1. Keep your box turned on its side because you are actually going to use the lid as a door.

2. The fan will be installed on the side toward the bottom of the box (elevated about an inch or two from the bottom).

3. Trace your fan on the outside of the box and cut out a hole.

PRO TIP: Once you cut your hole, it may be a good idea to tape over the exposed cardboard edges.

4. The fan should fit nicely into the hole.

INSTALLING YOUR HEAT LAMP

1. Now it's time to add the heat source for dehydration.

TRY ADDING A DIMMER: To be able to turn your lamp on and off and adjust the power, you can add a rotating dimmer switch to the middle of the power cord. The positive and negative wires in the dimmer switch will connect to the power cord running in through one side and out the other. Solder them in line with the light switch.

2. Run the wire of the heat lamp through the side of the box opposite the fan.

3. Cut a hole just large enough for the cord to fit through.

4. Your lamp should now function on a dimmable circuit.

ALSO TRY: Build a stand for your lamp so the bulb doesn't rest directly on the aluminum. Bend a hanger into a double "M" shape so that the lamp hovers above the bottom of the box to allow circulation and prevent overheating.

FOOD, FINALLY!

1. Now that we have it mostly assembled, it's time to add our cooling racks for the food.

PRO TIP: Buy your cooling racks before you choose a cardboard box so that you can select your box size accordingly.

2. You can put in a couple of cooling racks spaced a few inches apart (think of an oven).

3. Make holes in the sides at the top of the box and insert the wooden dowels. The dowels are for the racks to rest on. They should stick out of the box a bit so that your racks will stay balanced on top.

4. Put your lid on as the door, taping down one side to function as a hinge.

SMALL IMPROVEMENT: Cut several tiny holes in the top of the box to help the air come in one side, travel up past the food, carrying away any excess moisture, and out the top.

ALSO TRY: Build a little box for your dimmer switch so it stays connected to your dehydrator at all times.

5. Spray-paint and decorate to your liking.

NOW IT IS COMPLETE AND TIME TO GET COOKING!

Sweet, salty, or savory? There are so many recipes to try! Sugar-coated pineapple, apple slices, and beef jerky are just a few examples of the delicacies your dehydrator can cook up. Why buy dried snacks at the store when this method is faster, easier, cheaper, and just a cool thing to construct?

FUN FACT: Dehydration is just one of the many methods of food preservation. Up until the nineteenth century, a common practice was to cure and salt foods in order to preserve them but this still often resulted in spoilage. Finally, canning and sealing food in airtight containers was discovered and the modern method of dehydrating foods came after.

Don't you just love experiments you can taste? This project uses dry ice and a few simple ingredients to get a delicious treat with an experimental twist!

CARBONATED ICE CREAM

51

SAFETY KEY:
+ Do not swallow dry ice

SKILL LEVEL:
EASY
INTERMEDIATE
ADVANCED

APPROXIMATE TIME:
1 hour

WHAT YOU'LL NEED:
+ Dry ice
+ 2 cups half and half
+ Mixing bowl
+ 1 teaspoon vanilla extract
+ 1/2 cup powdered sugar

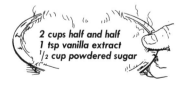

2 cups half and half
1 tsp vanilla extract
1/2 cup powdered sugar

LET'S BEGIN

SETUP

1. To make carbonated ice cream you will start by pouring 2 cups of half and half into a large mixing bowl.

2. Next pour 1 teaspoon of vanilla extract into that bowl.

3. Then add half a cup of powdered sugar and stir until it is evenly mixed.

ADDING THE DRY ICE: PUT THE "ICE" IN ICE CREAM!

1. Add ¼ cup of dry ice.

2. You will need to mix that around gently to cool down the liquid.

3. When you notice the fog stops flowing out of the bowl you will need to add more dry ice.

4. You will want to crush up the ice as much as possible. If you notice any chunks while stirring, pull them out with protective gloves or a ladle and crush them in a plastic bag.

WARNING:

☠ It is extremely dangerous to accidently swallow a piece of dry ice. This ice is -78 degrees Celsius and can leave you with internal frostbite and fill you up with carbon dioxide gas.

5. When your cream has the consistency of soft-serve ice cream, it is just about done.

PRO TIP: Inspect it one more time for hidden chunks and keep stirring until the vapor dissipates completely.

6. Once the vapor dissipates and there are no more dry ice chunks, it's time to scoop out your ice cream for a tingly treat!

CARBONATED ICE CREAM

NOT YOUR TYPICAL ICE CREAM!

While it looks like ice cream, there's a twist: It is fizzy. And now you know how to make a delicious-tasting homemade carbonated ice cream from four simple ingredients. Fizzy and delectable!

FUN FACT: A cow gives enough milk to make two gallons of ice cream a day.

If you have made it this far, congratulations! You are on your way to becoming a Master Tinkerer. This last project requires all the skill sets you have garnered throughout the year of weekend projects. It is now time to create an AK 47-style Handheld Rocket Rifle.

HANDHELD ROCKET RIFLE

52

SAFETY KEY:

+ Adult supervision: This project can be very dangerous due to the compressed air, so only use with adult supervision—and only outside and away from homes and buildings. Do not fire at pedestrians.

SKILL LEVEL:
EASY
INTERMEDIATE
ADVANCED

APPROXIMATE TIME:
6 hours

LET'S BEGIN

Because of the number of pieces needed for this project, we have helped you out if you want to order them from Home Depot online or at the store. We have included detailed part descriptions so you can have it all prepared.

Refer to the legend to see all the pieces and where they fit together.

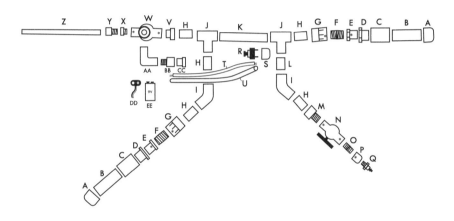

LABEL	QUANTITY	DESCRIPTION
A	2	2 in. PVC cap
B	2pcs × 9"	2 in. × 10 ft. PVC Schedule 40 DWV pipe
C	2	2 in. PVC Schedule 40 pressure slip x slip coupling
D	2	2 in. × 1 1/2 in. PVC Schedule 40 pressure spigot x slip bushing
E	2	1 1/2 in. × 1 in. PVC spigot x FIPT bushing
F	2	1 in. close PVC
G	2	1 in. PVC Schedule 40 pressure S x FPT female adapter
H	5pcs × 2.25"	1 in. × 10 ft. PVC Schedule 40 plain-end pipe
I	2	
J	2	1 in. PVC slip x slip x slip tee
K	1pc × 9"	Use leftover 1-inch pipe from "H"
L	1pc × 3"	Use leftover 1-inch pipe from "H"
M	1	3/4 in. × 1 in. PVC pressure S x MPT adapter
N	1	3/4 in. Sch. 40 PVC FPT x FPT threaded ball valve
O	1	3/4 in. x close PVC riser

LABEL	QUANTITY	DESCRIPTION
P	1	3/4 in. PVC threaded cap
Q	1	1/4 in. NPTM industrial plug
R	1	Gardner Bender 3/6 amp momentary-contact push-button switch
S	1	
T	1pc × 17"	Doorbell wire or similar insulated wire
U	1pc × 17"	5/16 in. OD x 3/16 in. ID x 20 ft. clear PVC tubing
V	1	1 in. PVC slip x MPT male adapter
W	1	Rain Bird 150 psi 1 in. in-line solenoid valve
X	1	1 in. × 1/2 in. PVC Schedule 40 pressure MIPT x FIPT Bushing model # 439-130HC store SKU # 796841
Y	1	1/2 in. PVC schedule 40 pressure MIPT x slip male
Z	1pc × 24"	1/2 in. × 10 ft. PVC schedule 40 plain end pipe
AA	1	1 1/4 in. PVC Schedule 40 pressure 90-degree slip x FPT female adapter elbow
BB	1	1 1/4 in. PVC schedule 40 pressure slip x MIPT male adapter
CC	1	1 1/4 in. PVC plug

ADDITIONAL ITEMS NEEDED:

From Amazon.com

DD	1	Fully insulated 9V battery snap connectors
EE	1	9-volt battery
Primer & Cement		Oatey 8 oz. Handy Pack Purple
Primer and Solvent Cement		
Air compressor		

OPTIONAL:

2–4 colors camouflage spray paint
1 can shellac finish and sealer

PVC PIPE ASSEMBLY

You can buy whole lengths of pipe, or just two-foot lengths of PVC pipe. Either will work.

1. 2" pipe: Cut two pieces to lengths of nine inches each.

2. 1" pipe: Cut five pieces to lengths of 2.25 inches each.

3. Cut one piece to a length of nine inches.

4. Cut one piece to a length of three inches.

5. Half-inch pipe: Cut one piece to a length of twenty-four inches.

At this point you should be able to "dry fit" the entire rocket launcher together to check fittings. Use the instructional diagram as depicted in the expanded-view diagram included herein. Remember that the "dry fit" is necessary before solidifying your launcher with glue.

WARNING:
☠ **Do not pressurize!**

NOTE: For cutting pipe, you can use a hacksaw or PVC cutting tool. Power tools are an option but can be dangerous and should not be used without adequate training.

GLUING TOGETHER

NOTE: Read and follow instructions on the labels of your purple primer and PVC cement cans.

1. Use purple primer on all surfaces that would be in contact, and let it dry for about ten minutes before applying PVC cement.

2. When all primer has dried, glue the parts together by applying a coating of PVC cement to both parts of the connection, and when connecting, twist them about a quarter-inch turn to ensure good coverage, and push fully together.

> **NOTE:** Only work on one connection at a time, and work quickly. PVC cement begins to set fairly rapidly, and any delay may result in a failed connection.

3. Glue all parts together as depicted in the expanded-view diagram. The threaded connections should not be glued.

4. Allow at least three hours for cement to cure before pressure testing.

ASSEMBLE

1. Drill a half-inch hole in the top of cap "P" for the pneumatic adapter.

2. You can tap the hole in cap "P" or just use the pneumatic plug "Q" to self-tap. Add thread tape to the threads, and screw down snugly. Do not overtighten because the connection may weaken and fail.

3. Add thread tape to any male-threaded connections (e.g., threaded risers) and screw together tightly with their associated parts. Make the connections airtight. Tools to help tighten may be required.

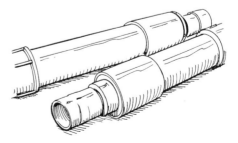

ELECTRICAL

1. Drill a ⁵⁄₁₆-inch hole in the side of cap "S." When attached, this hole will be on the bottom of the trigger. You may choose to grind down the bottom rim of the cap so it fits snugly on the handle of the launcher later when it gets glued on.

2. Also, drill a half-inch hole in the top of cap "S" just large enough for the button of the trigger switch to protrude from.

3. Drill a ⁵⁄₁₆-inch hole in the middle/top of end plug "CC."

4. Attach wires labeled "T" to the following: snap connector "EE," sprinkler valve "W," and momentary switch "R" as depicted here.

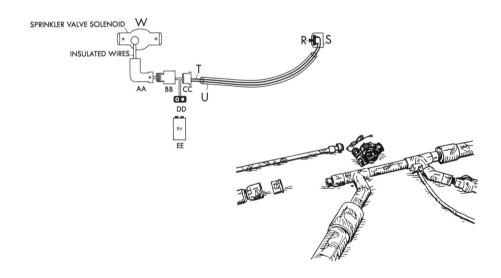

NOTE: To connect wires, the ends will need to be stripped of outer insulation and twisted together with the connecting wire, and any exposed wire needs to be covered with electrical tape to prevent electrical shorting on other parts of the system.

5. Connect nine-volt battery "DD" to snap connector "EE," and push trigger button "R" to test the connection. If connected properly, you will hear the sprinkler valve solenoid "W" clicking open and closed.

FINISHING

1. Pressure test the system. Attach an air hose from your air compressor and open the ball valve "N" to allow air to flow into the system. I recommend starting around 30 psi and slowly adding pressure, stopping to listen for leaks at regular intervals. Pressing the trigger switch "R" should open the valve and release all the air in the system. Do this on a regular basis to test the electrical connections. Do not exceed 135 psi.

2. As a final test you can close the ball valve "N," disconnect the air hose so that your launcher is portable, and test-fire in a safe area.

3. Hot-glue the trigger cap "S" to handle pieces "J" and "L" and allow five minutes to cool.

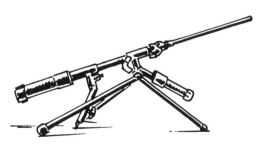

4. If painting the system is desired, use tape and/or masking paper to cover parts of the system you don't want painted, e.g., valve handles, pneumatic plug, sprinkler valve, trigger button, etc.

5. If possible, hang entire system from one single point, e.g., plug "Q," and spray paint as desired. Try a camouflage look or experiment with different colors and several coats of paint to give it the look you want.

6. Go above and beyond by adding a coat of finish to give the rocket launcher a professional sheen and help protect the paint.

7. Let dry for about two hours. Your system is finished!

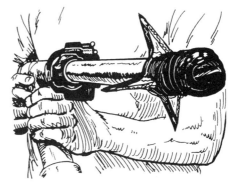

PAPER ROCKETS
(FOR YOUR HANDHELD ROCKET RIFLE)

SAFETY KEY:
+ Be careful using the hot glue gun, as the glue could burn you if touched.

SKILL LEVEL:
EASY
INTERMEDIATE
ADVANCED

APPROXIMATE TIME:
90 minutes

WHAT YOU'LL NEED:
+ Handheld Rocket Rifle (from chapter 52)
+ Hot glue gun
+ Electrical tape
+ 8½ × 11-inch piece of paper

+ Writing utensil (to mark the cardboard)
+ Cardboard
+ Scissors

LET'S BEGIN

THE BODY OF THE ROCKET

1. Cut the piece of paper down the center to make two 8½ x 5½-inch pieces. One half of the piece of paper will act as a spacer inside of the rocket (to be removed later) and the other half will form the rocket body.

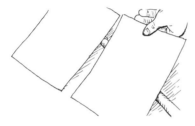

2. Roll the first half of the paper tight around the barrel of the Rocket Rifle and then tape it to itself. Wrap the second piece of paper around the first paper and tape it to itself, creating two paper layers that are just loose enough to be able to slide.

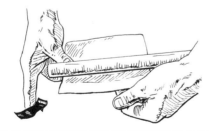

3. Move the top layer forward so that it hangs over the barrel of the Rocket Rifle by about one inch.

4. Fold the overhanging sides of the paper inward from four directions. Tape down the overlapping pieces of folded paper with electrical tape.

5. Wrap the pieces of paper in electrical tape starting from the base and work your way to the top in a circular fashion. Once at the top, wrap several layers of the tape over the top and around the sides of the top.

6. Smooth the tape out and wrap one more layer down the rocket body, ending where you first started wrapping.

ROCKET FINS

1. Get a piece of cardboard and use scissors to square up the edges.

2. Putting the piece of paper against the barrel of the rocket, mark the cardboard about a third of the way up the body of the barrel and then make a second mark at the top of the cardboard. The mark should be as wide as the barrel.

3. Use a straightedge to connect those two dots and cut the fin out.

4. Use this as a template to make three more fins (for a total of four) that are the same size.

5. Put the hot glue on the long horizontal side of the fin and then press it into place at the base of the wrapped barrel.

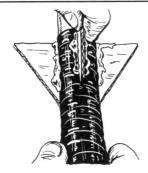

6. Once the glue has dried, repeat the process with the rest of the fins by rotating the barrel each time to make each of the fins symmetrical with the others around the barrel.

7. Use the electrical tape to wrap the tips of the fins twice. Smooth out the tape once it has been wrapped.

| FINISH |

1. Remove the inner layer of paper tubing. This will create a small gap between the inside of the rocket and the launcher, which will make it easy to load the paper rocket.

2. Add extra tape to the bottom of each fin to make them super durable.

3. Charge the air compressor to 135 psi. Once you've closed the ball, disconnect the air hose so that the air compressor (and your new paper rockets) are portable. Put the paper rocket on the top of the air compressor and shoot!

FUN FACT: The fins on a rocket are incredibly important. The fins keep the rocket stable as it flies through the air. This is because the fins are near the center of gravity of the rocket and keep it stable against the center of pressure.

IN MEMORY OF GRANT THOMPSON, 1980–2019

On July 29, 2019, Grant Thompson was killed in a paramotor accident in Southern Utah. Grant dedicated his life to learning, testing his curiosity, and pushing boundaries. He did it to benefit others and to share the joy of discovery. When this book was finished, Grant said, "I want the readers to challenge themselves and to make things in life they are proud of."

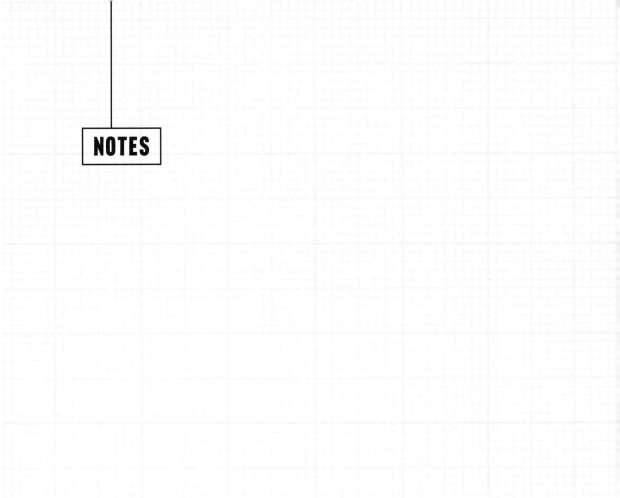

NOTES